BIODIVERSITY CONSERVATION

Editors

Dr. M.S. BINOJ KUMAR

and

Dr. P.K. GOPALKRISHNAN

Department of Post Graduate Studies and Research in Botany,

Sanatana Dharma College (University of Kerala),

Alappuzha - 688003, Kerala

SCIENTIFIC®
PUBLISHERS

Published by:
Scientific Publishers
5A, New Pali Road, P.O. Box 91
Jodhpur 342 001, India
sE-mail: info@scientificpub.com
Website: www.scientificpub.com

Print : 2025

ISBN: 978-81-72334-89-5 (Hardbound)
ISBN: 978-93-87991-13-2 (E-book)

Printed in India

PREFACE

Biodiversity encompasses all species of plants, animals and micro-organisms; the ecosystem and ecological processes of which they are part. Biodiversity includes genetic diversity, species diversity and ecosystem diversity. It is estimated that our earth supports about 50 million species of organisms, that have been the result of 3 billion years of evolution involving mutation, recombination and natural selection. The quantum leap in progress brought about by unprecedented advancement in technology which was responding to the ever increasing demands of a fast growing population has effected, as its rhythmic response, a level of exploitation that has almost literally pushed the natural resources to the wall. The present day life probably constitutes about 1% of the total number of life forms that have existed on this planet so far. With the explosive growth to human population, the life supporting system of earth is becoming increasingly threatened as the rate of global change accelerates. It is reasonable to suppose that by the end of the century we may well witness the elimination of about one million, out of the planets 50 million, species of organisms.

India is rich in Biodiversity as relevant both to the health of biosphere in general and to agriculture, animal husbandry, fisheries, forestry and pharmaceutical industry in particular. This diversity and crisis it faces are evidenced by the two hot spots on the globe, the Eastern Himalayas and Western Ghats consisting of 5332 endemic species of plants, mammals, reptiles, amphibians and butterflies. Expanding human population, overexploitation of natural resources, habitat loss and pollution are contributing to a greater extent to the rapid degradation of environment quality, with its irreversible loss of species diversity. The current perception therefore is that, if we fail to frame strategies to conserve biodiversity the future generations will blame us for robbing them for their life support systems. Conservation has been defined as the management of natural resources for the benefit of all life including human kind of the biosphere so that it may yield sustainable benefit to the present generation while maintaining its potential to meet the need and aspirations of future generations.

The papers included in this book were presented at the National Seminar on Biodiversity Conservation, organized by the Department of Post Graduate Studies and Research in Botany, Sanatana Dharma College, Alappuzha, Kerala from 8 to 10th October 2004. One of the most encouraging aspects of the seminar was the variety of papers presented from various fields of Biodiversity. This lead us to believe that the seminar met a real need and to hope that the proceedings will be useful to students and researchers engaged in diverse areas of Biodiversity Conservation.

We would like to express our sincere thanks to all the contributors for their cooperation in submitting the manuscripts with in the prescribed time limit.

We are obliged to the unflinching support from Adv. C.K. Parameswara Panicker, Manager, S.D. College; Shri. J. Krishnan, Secretary, S.D.V. Trust; Prof. K.C. Venugopal (Principal, Retd.) Prof. M.V. Sudharsanan, Principal, S.D. College and faculty members in the department of Botany such as Dr. V.K. Kusumakumari, Dr. S. Remadevi, Dr. B. Sreedevi, Dr. K. Ravikumar, Dr. V.S. Hari Kumar and Dr. P.R. Unnikrishna Pillai for successful convening of the National Seminar and brining out the proceedings.

We express our deep sense of gratitude to Dr. M.P. Nayar, Director (Retd.) and Dr N.P. Balakrishnan Joint Director (Retd.), Botanical Survey of India, for the inspiration and constant encouragement.

The financial support extended by the University grants commission, Bangalore, Kerala State Council for Science, Technology and Environment, Thiruvananthapuram and S.D.V. Management, for the conduct of the seminar is thankfully acknowledged.

Editors

CONTENTS

HOTSPOTS AND WARM SPOTS OF BIODIVERSITY

M.P. Nayar

Former Director, Botanical Survey of India, Environmental Resources Research Centre, Thiruvananthapuram

Introduction

At present the most serious environmental problem facing human race is the irretrievable loss of biodiversity leading to extinction of plant and animal species and erosion of genetic resources (Ehrlich and Ehlrich, 1981; Myers, 1991). Identification of species rich areas and centres of variability of species are important for any conservation planning of biodiversity. At a time when biodiversity is fast dwindling due to habitat loss, conservationists are charting out roadmaps to protect maximum biodiversity in the minimum area. This is required because the largest concentration of biodiversity at the species level is seen in the tropical forests of developing countries which have high population growth leading to the loss of habitats. Besides, the land availability for protected areas in the tropical developing countries is also limited. In any conservation matrix planning, it is necessary to take into consideration endemic centres, fragile ecosystem, genetic resources, habitat diversity, edaphic heterogeneity, catchment areas, and altitudinal topography with microclimatic ecological niches. As an instrument of protection of biodiversity through state intervention protected areas have been established through legislative procedures by several nations. Since the establishment of the first national park at Yellowstone in USA in 1872, over 8641 protected areas have been created all over the world covering approximately 5% of Earth's land surface (McNeely, 1994). The five IUCN accepted Protected Areas Categories are: Nature reserves, National Parks, National Monuments, Habitat/Species Management Areas, Protected Landscape/Sea scape. It is estimated that another 40,000 protected areas of varied nature which do not qualify for the IUCN categorization are in the process of establishment (Mc Neely, 1994). If this is included the protected area network would cover 10% of Earth's surface. In the conservation scenario, some of the concepts which have become engines of conservation action plans are the result of historic decimation of ethnic groups and ethnic cultures living in wilderness due to aggressive colonizers and immigrants, loss of habitats and extinction of birds and animals due to hunting as sport, loss of natural heritage sites and monuments due to urbanization and finally extinction of species and their ecosystems caused due habitat loss and fragmentation of forests. The IUCN categorization of protected areas is based on the above perceptions.

Hotspot concept: The conceptualization of protecting largest number of species in the smallest areas led Myers (1988) to the species rich tropical forests with their high degree of endemism. According to Myers "Hotsots" are characterized by exceptional concentrations of species with high levels of endemism in the tropical forest ecosystems and they face high degree of threat. On the basis of species and endemism analysis, Myers initially identified ten hotspots in the tropical forests and they hold 34000 endemic species, 27 percent of all known tropical species. The ten hotspots identified are as follows: Madagascar, Atlantic costal, Brazil, Western Eucador, Colombian Chaco, Uplands of Western Amazonia, Eastern Himalayas, Peninsular Malaysia, Northern Borneo, Philippines and New Caledonia. On further assessment Myers (1990) expanded the hotspot concept to 18 areas which support 49,955 endemic species which is 28 per cent of earth's plant species. The additional tropical forest sites included are Southern Ivory Coast, The Eastern Arc Forest of Tanzania, The Western Ghats of India. Besides the tropical forests, Myers (1990) also included Mediterranean type areas where there are high degrees of endemism: The Cape Floristic Province of South Africa, Southwestern Australia, California Floristic Province and Central Chile. In totality all the 18 hot spot areas cover only an area 0.5 present of earth's land surface. For conservation strategy Myer's hotspot concept was idealistic that is protecting the largest number of species in the smallest identified endemic areas. Since endemism is the main criteria, Mittermeier, Myers and Thomsen (1998) further expanded hotspots beyond tropical forests and Mediterranean types so that they cover some of the endemic areas of arid or semi arid regions. The following are the twenty four Biodiversity hotspots (Mittermeier *et al.*, 1998) prioritized in descending order according to the number of plant endemics: Tropical Andes, Mediterranean basin, Madagascar and Indian Ocean Islands, Mesoamerican Forests, Caribbean Islands,Indo-Burma, Atlantic Forest Region, Philippines, Cape Floristic region of South Africa, Eastern Himalayas, Sundaland, Brazilian Cerrado, Southwestern Australia, Polynesia/Microesia, New Caledonia, Choco/ Daren/ W. Eucador, Western Ghats/ Srilanka, California Floristic Province, Succulent Karoo, New Zealand, Central Chile, Guinean Forests of W. Africa, Wallacea and Eastern Arc Mountains and Coastal Forests. The total plant species endemic to the above mentioned hotspots are 124,055 species which represent 45.94% of plant species and the areas of occupancy cover only 2% of land surface. It is mentioned (Mittermeier *et al.*, 1998) that atleast 75% of terrestrial animal species which are categorized as critically endangerd or endangered or vulnerable are also found within the above regions (Groombridge & Baillie, 1996).

"Warm spot areas" of agro-biodiversity

It is seen that most of the hotspot areas do not cover grasslands and woodlands which are the major centres of agrobiodiversity as they do not have large number of endemic species. The major constituents of agrobiodiversity are as follows: (i) the harvested crop varieties, progenitors of cultivated plants, livestock breeds and their non-domesticated species (ii) Non-harvested species including micro-organisms, pollinators which support agro-sylvo-ecosystems (iii) Non-harvested species in the environment which support functioning of food production ecosystem. The Agricultural biodiversity is the result of the interaction of culturally diverse ethnic groups over time and space on the predomesticates. It includes the variety and variability of plants, animals and microorganisms which help the functioning and sustaining the key functions of the agroecosystems including its structure and processes, assuring the food production and food security (FAO, 1998). On the basis of the above criteria, it is proposed to designate areas of predomesticates and associated weedy species with vast arry of varieties and variabilities in grasslands and woodlands, which have evolved under various environmental stresses and co-adapted through man's interventions as "Warm spot areas". The major criteria are the occurrence of genetic resources usful to man, heterogeneity of landscape, extreme climatic, edaphic and altitudinal stresses which usually function as driving forces for evolutionary adaptations in plants and animals. The Warm spot areas include some of the Vavilovian centres of cultivated plants, but not all the Vavilovian centres of cultivated plants are Warm spot areas.

Savannas and Woodlands: Warm spot areas

It is appropriate to conserve the rich endemic biodiversity in the hotspot areas. What is not focused in the process are the areas of grasslands (Savannas, Steppes, Prairies and Pampas) and woodlands with heterogeneity of habitats and altitudinal variations with ecological niches having a vast array of varieties, cultivars and weedy species. These ecobioregions hold vast plant and animal genetic resources evolved in time and space along with hominids over the last two million years. The driving force of adaptations of predomesticates are usually the pressures of environment and climatic stresses which were closely interlinked with the man's pastoral life.

Tectonic episodes also played an important role in changing the landscape due to upliftment and subsidence causing isolation, disjunctions and convergence.

Aridity driving force of human and agrodiversity evolution

Past evolutionary history of human beings may give insights into the importance of savannas and grasslands. It is considerd that but for the upliftment of mountain systems cutting across vertically north to south from Ethiopian massif to Kenya and Tanzania in East Africa and the resultant formation of Rift Valley that human beings would not have evolved from the stock of chimpanzees which still inhabits in the dense tropical forests of Congo, east of the Rift Valley. Eight million years ago, the tectonic upliftment resulted in the formation of massive barrier separating the eastern portion of Africa from the western tropical forests of Congo-Cameroon. Due to this upliftment some of the ancestors of the common stock of chimpanzees of the eastern Congo forests got separated from the main stock. Those land mass on the uplifted E. Africa, due to desiccation and aridity became savannas or grasslands. Man's ancestors arose from this common stock of chimpanzees, which got separated from the Congo forests. The environemtal stresses of aridity put evolutionary pressures on hominids to hunt and gather food from the dry savanna grasslands. Over time and space, hominids evolved into *Homo erectus* and *Homo habilis* and finally into *Homo sapiens.* Coppens (1994), who studied extensively the Omo river valley cuttings and escarpments in Ethiopia through analysis of pollen flora, mentioned the history of climatic changes from tropical forest ecosystem to grasslands and this was well etched in the composition of tree and grass pollen flora in different layers in the Omo valley. The ratio of the pollen of trees to the grass pollen equaled 0.4 at the bottom of the Omo valley and it was less than 0.01 at the top. At the bottom, the pollens excavated belonged to the species of humid tropics ie. *Celtis, Acalypha, Olea* and *Typha*. The saga of evolution of hominids was also well represented by *Australopithecus afarensis* on the lower strata, followed by *Australopithecus aethiopicus, A.boisei* and *Homo habilis* in the younger upper strata. Coppens (1994) mentioned that human beings are the product of aridity. Just as aridity is the driving force of human evolution, aridity accompanied by uncanopied grass lands, seasonality, and altitudinal variations due to tectonic upliftments supported by human ingenuity in the selection of elitic species are the driving forces of the evolution of predomesticates and cultivated plants.

Warm spot areas as supplementary conservation strategy

Though the Hotspot concept of Myers, 1988, 1990; Nayar, 1996 is one of the most important useful paradigm for the conservation of threatened biota, it is worth noting that most of the hotspot areas are situated in areas which are highly threatened due to fragmented ecosystems (tropical and Mediterranean types) and extreme human demographic pressures. In looking for supplementary strategies, it is prudent to look for areas, which are not covered under hotspot concept but rich in biodiversity resources *i.e.* cradles of agricultural biodiversity evolution in the grasslands and woodlands. Besides, these Warm spot areas have vast geographical space and they have negligible population pressure unlike the Hot spot areas.

It is also noteworthy that most of the "Warm spot areas" are not Vavilovian centres of origin of cultivated plants though there are overlapping zones. The Warm spot areas occur only in the temperate or semiarid regions unlike Hot spots. The Warm spot areas give emphasise to varieties and variabilities under ecological, environmental and biotic pressures while Hot spot concept gives emphacise to highest concentration of endemics. It is assumed that varieties and diversities seen in populations of genetic resources seen in Warm spot areas are due to the possible meeting of cross fertilizing populations promoted by geographical space without any isolating barriers.

Warm spot areas of the World

On the basis of Warm spot areas concept - areas of predomesticates and associated weedy species with vast array of varieties and variabilities in grasslands and woodlands, which have evolved under various environmental stresses and co-adapted through man's intervention in time and space, the following areas are designated as "Warm spot areas".

1. Manchurian – Mongolian area.
2. Central Asia- Teinshan Area
3. N.W. Himalaya – Hindukush area
4. Transcaucasian – Zagros area.
5. Russian – Ukrainian Steppes area.
6. Ethiopian – Somalian area
7. Senegal Lake Chad sub Saharan area
8. Angola – Zimbabve bushveldts area.
9. Boliva-Paraguay-Gran Chaco Pampas area.
10. Appalachian mountains and eastern USA area
11. Sonora Arc and Prairies area.
12. Rockies woodlands and Semi arid Arizona.
13. Eastern Ghats and Semi arid Aravalli ranges.
14. Great Dividing Range of Australia.

References

Coppens, Y. 1994. East side story of the origin of humankind. *Scientific American* May 1994: 62-69.

Ehrlich PR and AH Ehrlich 1981. *Extinction*, Random House, New York. FAO, 1998.

Sustaining Agricultural Biodiversity and Agro – ecosystem functions, opportunities, incentives, and approaches for the conservation and sustainable use of Agricultural biodiversity in agro-ecosystem and production system. Report of the FAO/CBD *Agricultural Biodiversity Workshop*, 2-4, December, 1998.

Groombridge, B and J Baillie, 1964 IUCN Red List of threatened animals. *IUCN Publication Unit*, Cambridge, United Kingdom.

McNeely, JA. 1994. Protected areas for the 21st century: working to provide benefits to society. *Biodiversity and Conservation* 3: 390-405.

Mittermeier, RA, N Myers and JB Thomsen, 1998. Biodiversity Hotspots and major tropical Wilderness areas: Approaches to setting conservation priorities. *Conservation Biology* 12:516-520.

Myers, N. 1988. Threated Biotas. "Hot Spots" in tropical forests. *The Environmentalist* 8(3): 187-208.

Myers, N. 1990. Biodiversity Challenge: Expanded Hot Spots Analysis. *The Environmentalist* 10(4) : 243-254.

Myers, N. 1991. *A Wealth of Wild Species,* West View, Boulder Co.

Nayar, M.P. 1996. *Hotspots of Endemic Plants of India, Nepal and Bhutan.* 1-256. Tropical Botanic Garden and Research Institute, Trivandrum.

BIOHERITAGE OF INDIA - THREATENED AND ENDANGERED

N.P. Balakrishnan

Joint Director (Retd.), Botanical Survey of India,
13-14, B.P.C. Nagar, Thondamuthur Road, Coimbatore - 641 046

Abstract

The term 'Biodiversity' first coined in 1985 has become widely misused in recent times. An alternative term 'Bioheritage' is a more appropriate term implying the importance and preciousness of the biological wealth of the world. Man has continuously depended on plants for their existence and the human civilization itself depends on plants. The richness of the plant wealth of India is our bioheritage. In India the flowering plants alone amounts to about 17000 species, of which about 6500 species are endemic to our country. It is our duty to conserve and preserve them. The various threats faced by our bioheritage have been briefly discussed.

Traditional knowledge among villagers and tribal people who pass them from generation to generation should be considered as rich indicators of the many uses of plants for the benefit of mankind. They should be well documented and patented before multinationals grab and patent them.

It is surprising that loss of plant wealth is not considered as major problem by politicians, bureaucrats, engineers and planners. Unfortunately in recent times the natural heritage around the world has become threatened and endangered. Developmental activities encroaching into the natural forest areas have depleted the plant wealth, and the pace of such activities is increasing day by day. Among the various threats to our natural heritage the most dominant ones are habitat loss for construction of dams, roads and factories, genome exploitation by pharmaceutical and other industries and consequent genetic erosion, timber extraction, ecotourism, hydro-carbon pollution of environment and its greenhouse effect and most important the population pressure. Unless we put control on such activities, most of our endemic species will become extinct before end of this century. The term “Biodiversity" was first coined in the year 1985 as a contraction of "biological diversity". Biodiversity is often defined as the variety of all forms of life, from genes to species, and through genera, families to the broad scale of ecosystems. This term has taken on a meaning and import all its own portraying the multiplicity of all biota, including animals and plants together, and the interaction among them in an ecosystem.

In 1988, biodiversity did not appear as a keyword in Biological Abstracts, and biological diversity appeared once. In 1993, biodiversity appeared seventy-two times, and biological diversity nineteen times". Today it is hard to count how many times the term "biodiversity" is used every day by scientists, policy-makers and others. While the history of this term is relatively short (as compared to other terms) it already has raised important, distinctive, philosophical issues. Overall, the major issue for biodiversity is how its conservation may be integrated with other needs of society.

Biodiversity, as used now, mostly deals with the number of taxa and most commonly the number of species. However, the term Biodiversity has much wider implication and meaning. It is a continuous phenomenon transgressing all hierarchical categories. It portrays variations of characters from family to individual plants. The biodiversity of a family continues into the genera, and then into the species, into populations and ultimately to individuals. It can be plant diversity, animal diversity, family diversity, species diversity, population diversity, ecological diversity or genetic diversity, all intergrading into one another. No demarcation is possible to decide where it begins and where it ends. In most of the recent papers claiming to be on biodiversity, it is plurality of plant resources that is depicted and not diversity. The degree of diversification is not adept for measurement and quantification. The degree of plurality is measurable by the number of elements or taxa within a category, whether it is a family, a genus, an area, a state or a country. In essence Biodiversity is too general a term, which refers to the variability among all living organisms, diversification among them, irrespective of the position or category. It is inappropriate to use it for depicting the number of elements in a domain. Its meaning differs between users depending on the purpose. Introduction of this term was not meant to replace the orthodox terms, like Fauna, Flora and Taxonomy. There has been much misuse of the term Biodiversity, often applying it to fields which do not reflect the real diversity of the organisms. I argue, that those papers which enumerate the taxa present in an area cannot be called Biodiversity of the area, but only as bioplurality, flora or taxonomy of that area. However, since it is useful to have a term to depict the entire biota of an area in its totality to understand its richness and uniqueness, I present here an alternative term to replace Biodiversity.

After much thought, I introduce the term **Bioheritage** as the most appropriate one to visualize and understand the multitude of organisms, their relevance and importance to the human needs. Biology is the study of living things and their vital processes. Biota is the term used collectively for all living things on earth. Heritage means a property that descends from generation to generation, something transmitted by or acquired from a predecessor or something possessed as a result of one's natural situation of birth. Bioheritage visualizes and amalgamates all the Biota of an area with their accompanying biodiversity as well as biounity. It depicts the richness and preciousness of the biota which we have inherited through millions of years of evolution and reminds us that it is irreplaceable and non-duplicable, warranting love, respect, protection and preservation for posterity. Essentially Bioheritage consists of the wealth of life on earth, the millions of plants, animals and micro-organisms, including their genetic diversity, species diversity and the intricate ecosystem diversity.

This term goes hand in hand with the other heritages, i.e. the cultural heritage, religious heritage, architectural heritage, musical heritage, linguistic heritage, etc. It also reminds us that we are not owners of the biota but only caretakers whose responsibility is to maintain it and hand it over to our children and grandchildren. It also implies that we should hand over the biota to our next generation in almost the same condition as we got it from our parents and grandparents.

Therefore, I think Bioheritage is a better general term than Biodiversity.

Civilization depends on plants

Somebody has said that civilization was built on food-grains as much as on bricks and mortar. We have an astounding whole-grain heritage starting about 20,000 to 30,000 years ago. Ehud Weiss of Harvard University and his team came across more than 90,000 plant remains when they excavated a Stone Age settlement on the Sea of Galilee in Israel. Cereal grains stopped our ancestors from spending all their time gathering food, and gave them time to build up their art and culture.

Plant wealth is of prime importance for the Earth's biosphere. With the exception of some sulphur bacteria, all animals are ultimately dependent on the photosynthetic plants. Man has continuously depended on plants for their existence, as well as cultural, scientific and technological advancements. Thus plants and animals form our bioheritage, the inseparable part of our very existence. We have a responsibility to preserve and protect it for our own self preservation and perpetuation. It would not be an exaggeration if we say that humans are part of all living things and all living things are part of human beings. Man – despite his artistic pretensions, his sophistication, and his many accomplishments – owes his existence to a six inch layer of topsoil, the rains and the grains.

This Bioheritage at present is becoming threatened and endangered due to various activities of man. Now let us examine what are the major threats and how to reduce those threats so that what we have today as our Bioheritage can be preserved and protected for the benefit future generations.

Areas with concentration of Bioheritage

When we look at the various climatic areas on earth, we find that Bioheritage is rich in rain forests, mountains, arid and semiarid areas, humid areas, primarily due to diversity in soil, climate and other physical structures. Tropical rainforests, where it rains on more than 200 days in a year, are the richest in bioheritage. Almost all such forests lie near the equator and are home to over 50 per cent of the world's species and plants and animals. Such areas are considered as hot spots, meaning that they are untouchable and should be left to be observed and studied, not for interfering with it and or meddling with its intricate balanced equilibrium.

India, the seventh largest country in the world, is a land of great natural diversity. India is one of the world's 12 megadiversity countries possessing a biogeography which is distinctively unique on the earth. With only 2.4 per cent of the global land area, India possesses about 11% of the world's biota. There are more than 130,000 species of living organisms representing 45,000 plant species and about 85,000 animal species in India. Among the plants there are about 17,000 flowering plant species and out of which almost 6500 species are endemic to our country. This is our bioheritage and it is high time that we realise its importance and remind ourselves that it is our duty to protect and preserve them for posterity.

Forests silently perform innumerable ecological services to mankind. They assist in global cycling of water, oxygen, carbon and nitrogen. They lend stability to hydrological systems, reducing severity of floods and permitting the recharge of springs, streams and underground water table. Trees prevent soil erosion from hills and mountains; keep sedimentation of rivers and lakes and also house millions of plant and animal species. They act as carbon-sink, in the sense that the carbon dioxide is absorbed and its accumulation of the atmosphere is reduced.

The rapid depletion of natural flora and fauna due to habitat destruction, deforestation for timber extraction, dam construction, road building etc. and other human activities has caused considerable depletion of our rare forest species many of which are endemic species with restricted distribution. Many of them have already become extinct.

Threats facing Bioheritage

1. Habitat loss

While there are many types of threats to biodiversity in this country, destruction of habitat has become the most significant. This habitat loss takes a number of forms, but much of it is the direct result of urbanization, as when a wetland is filled for a new home or a forest is converted into a monoculture plantation. Urbanization patterns and trends create serious problems of habitat fragmentation, when habitats become splintered by roads and other forms of development. This habitat loss is the result of wasteful patterns of high-density development. These threats place bioheritage loss squarely and undeniably in the realm of planning. We are building huge dams across our mighty rivers to produce electrical energy and to irrigate our cultivated lands. Unfortunately such dams when constructed inside dense forests submerge vast areas of natural forests and the biological wealth of that area suddenly become extinct and can never be retrieved again.

Current deforestation rates have been conservatively estimated by the United Nations at about 30 hectares per minute or 15 - 20 million hectares annually. They say that we would be losing on an average 5000 species per year over the next several decades. These losses would include about 20 species of mammals, 40 species of birds, and 1,250 species of plants per year (Lakshmanan *et al.,* 1998). Most of the tropical forests have been wiped out, even before their importance to mankind was realized. Only a minor belt now remains in South America, Tropical Africa and Indo-malesian area.

In Kerala only 7% of the land area is occupied by real virgin forests. The present situation has come to about due to continuous encroachment of primary forests in Western Ghats for cultivation of plantation crops, illegal settlements for human habitats and cultivation of crops and construction of hydro-electric dams. The Silent Valley hydro-electric project, which was stalled about 20 years ago by concerted efforts of environmentalists and biologists, is being revived under a new name called Pathrakadavu Hydro-electric Project. There is also talk about the revival of the Athirapally hydro-electric project, which was stalled after much public debate about its harmful effect on the virgin forests of the area. I appeal to all enlightened environmentalists and biologists of Kerala to put up a joint fight to stop these projects. My one and only argument against these projects is a question to the Engineers, Politicians and Bureaucrats. While there are several other ways for producing electricity with all your skills and modern technological advances, is it possible for you to replicate a natural tropical rain forest with all its original diverse biological entities? It is humanly impossible to recreate what God has given us as Bioheritage, and you have no right to destroy it.

Forest Department claims that the forest area of the country has increased during the last two decades. Such increase is only in plantation forestry and not natural forests. What to a layperson is an intact forest may be botanically depauperate. It is doubtful if conventional conservation strategies, i.e. declaring selected areas as sanctuaries, national parks, biosphere reserve, botanic gardens, gene banks, etc. can be really effective to cope with future needs of humanity.

As we transform the world's landscape by cutting down forests, draining wetlands, building roads and dams, there is a new danger of new diseases arising and posing threat to humans. In the July, 2004 issue of *Environmental Health Perspectives*, a team of experts report that widespread changes of global landscape are opening up new avenues for dozens of infectious diseases including malaria, dengue fever, cholera, influenza, foot and mouth, hemorrhage fevers. The opening up of a closed ecosystem like the primary forest releases many pathogenic microorganisms into human habitats.

2. Eco-tourism

Tourism is now India's largest industry, providing 15% in foreign exchange earnings, and wildlife tourism is its fastest growing new concept. With the potential to deliver massive economic benefits, global tourism has also been found to contribute to environmental degradation and to the undermining of human rights. Whilst there is need to harness the positive potential of tourism in India, care must be taken to manage tourism in a manner that does not compromise the integrity of natural resources, including forests and wildlife.

Tourism has been considered a major issue by environmentalists so far as environmental damage to natural tourist sites is concerned. Poor planning of tourism schemes and alarming growth of the industry are major factors responsible for the damage. There is no limit on the number of tourists but there is a limit for what the environment can tolerate and bear. Mass tourism brings with it hordes of people wanting to make quick money out of it without the much needed sensitivity. Hotels and lodges are constructed near the shores, coasts of lakes and estuaries. The refuse leftovers and sewage water get into the waters polluting it and damaging the aquatic ecosystem and depleting fishing resources of local people. Jetties are constructed for the boats. Country boats are replaced by mechanised boats which not only create a lot of sound but also leave traces of oil/diesel/petrol into the waters. Besides insensitive tourists may damage the rare flora and fauna of the ecosystem. All these problems are faced by Dal Lake (Srinagar) or Chilka Lake (Orissa) and by almost all wetlands around the world. Eco-tourism is being promoted in Kerala on a large scale by private business. Very soon problems will be felt along the inland waterways of Kerala, if sufficient precautions are not taken now itself.

Proper attention should be paid to develop tourism only in conformity with the environmental requirements and must merge harmoniously with the overall action plan for conservation and development. It is fashionable these days to add the prefix "eco-" to anything that needs rejuvenation or modernisation. The application of the term `eco-tourism' has often caused controversy, mainly due to the fact that some providers of conventional tourism use the concept as a marketing tool, without embracing its ethical and environmental basis. The World Conservation Union (IUCN) adopted the following definition of ecotourism:

"Environmentally responsible travel and visitation to relatively undisturbed areas, in order to enjoy, study and appreciate nature (and any accompanying cultural features – both past and present) that promotes conservation, has low visitor impact, and provides for beneficially active socio-economic involvement of local populations."

The above definition demonstrates ethical and holistic approach of ecotourism, in which the well-being of local people, conservation of natural areas and the running of a profitable business are taken as a set of integrated management goals. Ecotourism operations typically tend to be small-scale and sensitive in terms of their impacts on the local environment and culture, involving the participation of local communities as major stakeholders.

When properly managed, such ventures can result in significant benefits to the local community, through increasing employment opportunities, diversifying the local economy and providing funds for conservation. The ecotourists themselves have a more fulfilling experience by experiencing natural areas in a manner that raises their awareness of sustainable development and engenders respect for local cultures.

Environmental degradation can lead to a decline in tourism. Uncontrolled tourism leads to environmental degradation.

3. Gene piracy

The economic significance of Indian bioheritage can be gauged by the fact that the domestic trade in medicinal herbs and its extracts is to the tune of Rs. 3 billion per year and it is increasing.

Access to bioheritage should be distinguished from access to genetic resources, despite the difficulty to draw the line between both categories. This is because genetic access has never been regulated and genetic resources had been considered as the common heritage available to everybody in the world. Further, the monetary gains arising out of the use of genetic resources are significantly higher than those arising from physical access.

Europe and North America, during their formative years in the 19th century, had their economy rooted in agriculture. The transition from agricultural economy to industrial economy started only during the 20th century, more particularly after the Second World War. Today only 5-7 per cent of their population is engaged in agriculture, while it is as high as 60-70 per cent in India. Therefore, Agriculture will remain central to our development well throughout the current century. In this background, seed laws become important, lest we have another recurrent Texbasmati or Turmeric episode. The recent American efforts to patent Turmeric, Basmati rice, Neem etc. should open our eyes. India is no match with technologically advanced nations with capital resources in doing agro business. The giant corporations of these nations are precisely targeting our rich gene bank. They are well-equipped in sophisticated recombinant DNA technology. During the last few decades our rich gene bank was callously raided by these gene pirates. With globalization and privatization, the task is becoming easier. Our genes embedded in the seeds are our most precious treasure and we need to guard them at all costs. We should not allow them to be pirated and utilized by Multinational companies who can utilize them for producing Genetically Modified (GM) crops.

GM technology permits companies to ensure that everything we eat is owned by them. They make sure that crops cannot be grown without their patented seeds. This technology can also ensure that seeds of the GM plants we cultivate cannot reproduce themselves. By buying up competing seed companies and closing them down, they can capture the food market. Thus GM foods enforce dependence on monopoly seeds and monopoly pesticides.

My strong conviction is that GM crops are not science; they are technological products of science. Multinational companies spending huge amounts of money in GM crops are not doing it for scientific progress but only for enriching their coffers. We Indians are highly gullible and I warn that we should not fall into their trap.

4. Environmental pollution

Everybody knows that fossil fuels are the main polluting energy resource. The naturally occurring greenhouse gases such as oxygen, nitrogen, carbon dioxide, etc. keep the ground temperature at normal level. The greenhouse gases trap incoming solar radiation and thereby keep the Earth's surface warm. Burning fossil fuels such as coal and petroleum products liberate more carbon dioxide into the atmosphere.

Other activities like raising cattle and planting rice emit methane and nitrous oxide. Industrial activities release carbon into the atmosphere. All these make greenhouse gas blanket thicker and result in increase in global temperature. Scientists of Princeton University estimated that the average global temperature would rise by 2°C in the 21st century.

The United States Environmental Protection Agency (EPA) has found that the air in India shows very high levels of airborne particles ejected from vehicles and industries, which will soon produce dangerous greenhouse effects in the subcontinent. Therefore it suggests that efforts should be made immediately to reduce the emissions of greenhouse gases to the level below what was available in the year 1990, for which at least a 5% reduction in emissions is urgently needed. It is time to save and keep the Earth clean and green by reducing pollution causing activities in order to save the bioheritage.

Deforestation destroys carbon sinks. A vast amount of carbon is stored in the extensive forests of the tropics, particularly in the massive trees of virgin evergreen forests. The release of that carbon through deforestation and burning could add significantly to the atmospheric carbon dioxide.

Traditional knowledge

India is home to many different races and cultures. Until perhaps a century ago, due to the absence of geographic or economic integration, every group more or less continued to live as their forefathers in relative isolation. One indirect benefit of this isolation has been in the field of traditional knowledge about the many uses of local plants and animals among the many tribal and village communities, which continued to flourish and pass through generations. Many tribal people and villages near forests lost their traditional knowledge as a result of economic and geographical integration with the so-called modern civilization. However, there still remains vast amount knowledge in tribes and villages living in remote areas.

A traditional village healer may not reveal his or her knowledge lest it loses its significance on being told. This belief may seem unscientific but at the same time ensures that a complex or risky recipe is not pursued or practiced by someone untrained in the art. It is not my contention to argue that traditional knowledge systems are better than institutional knowledge. In many cases, when we take a plant or some other element of local knowledge systems out of its institutional context, even if a scientific relationship between cause and effect does not get adversely affected, the *institutional* context in which the plant is collected (for example, only when necessary and only in limited quantities) may get affected. Therefore, we may be able to develop a good and effective drug by just dealing with the utilitarian part of the traditional knowledge systems. But we may not necessarily maintain the restraint that may have been kept in place by some of the traditional institutions for conservation of that plant. That is the reason why many groups oppose bio-prospecting by outsiders in order to avoid the risk of over-exploitation of the resource itself.

The local beliefs in the power of spiritual icons have helped conserve sacred groves, lakes, mountains, etc. all over the world. The sacred beliefs are linked sometimes to very basic functional needs. For example, the need to protect the sandy mouth of the rivers is actually meant to prevent sea water going upstream and damaging the crops along the riverbanks. The points at which rivers originate are considered sacred almost all over the world, because that area ensures the slow flow of the river water through rich vegetation. The rate of erosion of local knowledge about bioheritage has never been as high as it is in the current generation in areas.

We prevent local tribal people, village healers and *vaidyas* access to plants and animals in a Biosphere Reserve, National Park or protected forests, but at the same time foreign plant hunters from international institutions and MNCs with money power and local institutional support finds it easy to enter such areas and collect what they want. Access to bioheritage for local communities to ensure their sustainable livelihood systems, should take priority over access for outside institutions or individuals.

Agro-bioheritage

Ever since human race began, man depended mainly on plants for all his essential needs of food, clothing, shelter, medicines and a host of other uses and approximately 80,000 species of plants have been used at one time or other since then. About 3000 species of plants have been used somewhat consistently since the beginning of agriculture. However among them only about 150 species have been cultivated on a large scale. Fewer than 25 species of crops supply more than 90 percent of the world's food needs (Ayensu, 1983, Harlan, 1975). There are innumerable primitive cultivars and land races among the traditional agriculture being practiced by tribal people in our forests. These races with rich diversity in genetic resources are rapidly getting lost by the introduction of high yielding varieties. There is urgent need to identify and protect them for use in future breeding programmes.

Agro-bioheritage primarily refers to genetic diversity in cultivated plants together with their closely related wild plants. Many traits incorporated into modern crop varieties were introduced from wild relatives. Therefore the wild relatives of cultivated plants are of great importance for improving the performances of our cultivated plants. According to Zeven and Zhukovsky (1975) the genetic diversity in food plants is concentrated in 12 major regions, of which one is India. There are numerous wild relatives of cultivated plants in our forests.

Western Ghats has been considered as one of the two areas (the other being Eastern Himalayas) rich in wild relatives of cultivated plants, called Hotspots. For example the estimate prepared after detailed field studies by Mohanan and Sivadasan (1999) in a small portion (ca 200 sq. km) of the Agastyamalai range at the southern-most end of Western Ghats, showed that the area has 93 species of wild relatives of cultivated plants and among them 33% are endemic to the area. The various categories are 11 species of cereals and millets, 3 species of pulses, 7 species of vegetables, 8 species of tuber crops, 39 species of fruit crops, 22 species of spices and 3 species of oil seed plants.

In situ conservation of agricultural bioheritage can be accomplished by the maintenance of the various cultivars, races and the genetic diversity present in and among populations of the many species used directly in agriculture, or used as sources of genes, in habitats where such diversity arose and continues to grow. It concerns entire agro-ecosystems, including immediately useful species (such as food crops, forages, and agro-forestry species), as well as their wild and weedy relatives that may be growing in nearby areas.

Over time, natural interbreeding between crops and wild relatives added to crop diversity and health. In modern varieties of potatoes, wheat, barley, rice, maize and oat, among other crops, traits from wild relatives have improved productivity as well as tolerance to pests, disease and difficult growing conditions. Many wild crop relatives continue to exist only in centers of domestication, predominantly in the developing countries.

It has been reported by Helmut Habel of Institute of Interdisciplinary Research, Austria that very soon the humanity would be living beyond its means. Using 'ecological footprint' they calculated the area of land needed to produce the energy needed for each individual to maintain the present level of consumership. They found that worldwide ecological footprint was already 20 per cent larger than the area of land available. In Austria, with about 8 million people, 5.4 hectares was needed for each inhabitant, whereas only 4 hectares were available. Each U.S. citizen needs ten hectares of land to cover his or her present day level of consumption. India, with a land area of 329 million hectares caters to the needs of 1000 million people and the per capita land available for each Indian is 0.3 hectares! The population of the world is currently about 6.4 billion and it is increasing by about 40 million per hear or 192,000 per day. The world average at present is only two hectares of land per inhabitant. If the entire humanity reaches the U.S. level of consumption, humanity would be living beyond its means.

Yet human beings must be fed, clothed, and housed. We often state that the three basic needs of man are food, clothing and shelter. However, we often forget that we need plants in some way or other for all these basic requirements. Each one of us is consuming 20 per cent more than what the planets bio-productive area can provide, meaning that we may need a whole new earth in about 50 years! Every day we do things and use things that deplete the bioheritage of the earth and cause pollution which harm the entire living world.

We all moan and groan about the loss of the quality of life through the destruction of our ecology, and yet each of us in our own little comfortable ways, contribute daily to that destruction. It is time now to awaken in each one of us the respect and attention our beloved mother earth deserves.

We should develop awareness among people about the value and importance of our bioheritage. Research in bioheritage, conservation and management activities must not only be directed for the good of the plants and animals, but must also be directed to the people living alongside. It is high time that we declare that the Bioheritage of India is our national wealth and not open to international commercial institutions. We should immediately enact all types of laws that are necessary to protect this irreplaceable wealth form destruction, encroachment and greedy piracy.

References

Aayensu, E.S. 1983. The world's diminishing plant resources. In: S.K. Jain and K.L. Mehra (eds.) *Conservation of tropical plant resources*. Botanical Survey of India, Howrah.

Harlan, J.R. 1975. *Crops and man.* American Society of Agronomy, Wisconsin.

Lakshmanan, D. Guruswamy and Jeffrey A. Mcneely 1998. *Protection of global biodiversity: Converging Strategies*. Duke University Press

Mohanan, N. and M. Sivadasan 1999. Crop-plant relatives of Agastyamala, Western Ghats, Kerala. In: M. Sivadasan and Philipp Mathew (eds.). *Biodiversity, Taxonomy and Conservation of Flowering Plants,* Pp. 27 – 34. Mentor Books, Calicut.

Zeven, A.C. and P.M. Zhukovsky 1975. *Dictionary of cultivated plants and their centers of diversity*, Wageningen.

PEOPLE'S BIODIVERSITY REGISTER: A PARTICIPATORY PROJECT CASE STUDY OF ERNAKULAM DISTRICT, KERALA STATE, INDIA

M.K. Prasad

62, Girinagar, Cochin - 682 020, India

Introduction

An important perspective emerged after the Convention on Biological Diversity (CBD) developed at the Earth Summit in Rio de Janeiro in 1992. CBD accepts the sovereign rights of countries over their biodiversity resources: acknowledges the neeed to respect the knowledge of sustainable use of living resources by local communities and to share with them the benefits of the commercial application of such knowledge. Such local community knowledge needs therefore to be documented, not anonymously as in the normal ethno-biological literature, but with specific acknowledgemnt of persons or communities involved.

Peoples' Biodiversity Register (PBR) is a programme of collecting information available in the domain of flock science on the status, uses and management of the biological resources. Such information is not documented, and is almost always orally transmitted to later generations. Many a time they are guarded as secrets and are likely to be lost if the recipient generation is not keen in preserving them. Chances are that such information will disapper when people loose control over the local resource base or when they have access to newer resources in their day to day life. In due course of time the PBRs may attempt to collate, verify and make the information available to the society at large, as well as feed it back to the village communities.

Objectives

The main objectives of the programme consists of creating decentralized database on :

1. Status of biodiversity resources like localities and population size of medicinal plants, cultivars and their wild relatives, vegetables, fruit tresss, ornamental plants, cash crops, timber trees, fodder crops, domesticated animals, animals of medicinal importance, fishes and pests.
2. Various factors such as harvests, chages in agricultural practices.
3. Involvement of local individuals/communities in sustainable use and conservation of biodiversity
4. Local knowledge widely shared and publicly devolved of properties and uses of biodiversity resources eg., use of certain medicinal plants in treatment of human or animal ailments.
5. Local knowledge only partially disclosed eg.a claim that a particular medicine – man/woman knows a cure for snakebite.
6. Perceptions of local people on participatory management options.
7. Create a network of decentralized database giving full credit to the origin of information at individual or community level.
8. Provide information such as volume of trade and prices of economically important elements of biodiversity.
9. To protect local biodiversity and local knowledge from being privatized by commercial interests who may file patents on modified products, processes, and biomaterials that are built up on local resources and knowledge.

Materials and Methods

The PBR project was sponsored and supported financially by the Ernakulam District Panchayath (the third tier of the local government under the democratic decentralized governance in the state of Kerala) during 1997-99. The methodology was adopted from the standardized

People' Biodiversity Programme of the Centre for Ecological Sciences, Indian Institute of Science, Bangalore.

The methodology for field investigations included the following components:

1. Building rapport with local people.
2. Identification of different used groups
3. Identification of resource persons and volunteers and their training.
4. Identification of village elders as sources of information.
5. Interviews with individuals as well as members of user groups.
6. Mapping the landscape of the study area.
7. Discussions with village assembly.
8. Field visits for truth testing if reqired.
9. Discussion with outsiders affecting resource use at the study site.
10. Documenting knowledge and perceptions of local community.

The programme consisted of training local volunteers, mostly students and unemployed youth, in the technique of gathering information based on a questionnaire, through informal dialogues with selected village elders. Such information was scrutinized by resource persons, consolidated, and recorded in specially designed data sheets, which were bound into books. The register also contains information on the history, geography, demography, ecology etc. of the village involved. About 10,000 people participated in the whole programme.

Results

1. User groups

Composition of the human population in the study area which is a collection of 86 Gramapanchayaths (the first bottom tier in the local government consisting of one/two or more villages having a population of 25000 approx.) was documented. Communities that earn their livelihood by traditional occupations like the Potters, Weavers, Basket makers, Mat weavers, Blacksmiths, were listed. It was noted that such groups are fast disappearing. The focus of their listing was with reference to their relationship with bioresources.

2. Landscape

The second step involved in the mapping and description of the relevant landscape, both in terms of vegetation cover with use of appropriate terms of landscape ecology of the locals. These maps form an important tool in understanding the patterns of use and managements of the different landscape elements. The strategies for conservation and sustainable use of the biological resources of the local communities are spelt out on the basis of this.

3. History

The history of the user groups, the landscape and the biological resources of the study area was also recorded. Such history is important to an understanding of the current status, use patterns and management options for the future.

4. Knowledge base

Recording of the knowledge of the local people of the landscape, the biological resources of the present and past, and how they managed them was the next step. This coverd names and uses of 1023 important living organisms use of certain plant parts in treatment snake bites, pediatric illness etc, which knowledge is kept as secret and transmitted from generation to generation orally.

5. Bioresources

Another module of the PBR is an inventory of the current occurrence and history of the biological resources as recounted for people for different elements of the study area. This inventory contains 310 species of medicinal plants, 43 vegetables, 42 grasses, 109 ornamental plants, 129 timber/fruits trees, 42 wild relatives of cultivars, 38 chewing/incense plants, 44 animal species of medicinal importance, 95 species of fishes and 51 pests, 314 varieties of rice were cultivated once.

Economic utilization

The PBR also contains a section on the various uses of bio – resources for cash incomes. There have been instances of non-sustainable use of some resources inspired by market forces, subsistence cultivation giving way to cash crops and fishing for local consumption giving way to high-tech aquaculture.

Conservation measures

Traditional protection to sacred groves, sacred species like Peepal tree are some examples of conservation measures met with. Self-imposed regulation of fishing in the neighbourhood wetlands is another example.

Development aspirations

The development aspirations of local people arising out of interviews were also documented. These are useful in designing local level planning for development and possible conservation and sustainable use of bio-resources.

Management options

The priorities of local people with regard to conservation of different elements of landscape and different species of plants and animals and the strategies they adopted in the past are also listed.

Acknowledgement

The author wishes to acknowledge the academic and management assistance of all the Resource Persons, volunteers, elected representatives of the Panchayaths and the Ernakulam District Panchayath authorities in this programme. Special thanks to hundreds of village inhabitants and the village elders, without whose cooperation this project would not have been possible.

Reference

Madhav, Gadgil, 1996. New Meanings for Old Knowledge: The Peoples' Biodiversity register programme" Centre for Ecological Sciences, Indian Institute of Science, Bangalore.

Achar, K.P. 1997. Documentation of people's knowledge and perceptions about Biodiversity through PBR at Mala Village Panchayath, Karkala. Karnataka State.

ECOLOGY AND BIORESOURCES OF THE WETLAND ECOSYSTEMS ON THE SOUTH WEST COAST OF INDIA WITH SPECIAL EMPHASIS TO THE RAMSAR SITES

S. Bijoy Nandan

Central Inland Fisheries Research Institute, Alappuzha Centre,
Alappuzha- 688 001, Kerala

Abstract

The South West (S.W.) coast of India is blessed with a series of unique wetland systems, the backwaters, lagoons, inlets, barrier islands, popularly referred to as *Kayals* in this part of the country. The wetland systems are preferred habitats for over 200 resident or migratory fish and shellfish species, and the fishing activities in these wetlands support livelihood to about 0.2 million fishermen and also provide full-time employment to over 50,000 fishermen in the region. This contribution highlights the environmental and fishery investigations in 12 of the 30 wetlands, undertaken during 1994-2002 period. The pH was generally near neutral to alkaline in range. The salinity values indicated mixohaline condition ranging from 5.20-32.38 ppt. in the 12 wetlands. The productivity values were generally low in most of the wetlands during the study, where the gross production varied from 0.22 $gC/m^3/day$ in Kadinamkulam to 1.10 $gC/m^3/day$ in the Kayamkulam. The diversity of plankton and benthos was more during the pre-monsoon compared to the monsoon and post-monsoon periods in most of the wetlands. The average fish yield per ha. varied from 246 kg. in Valapattanam to

2747.3 kg. in Azhikode wetland. Retting of coconut husk in most of the wetlands led to acidic pH conditions with anoxia resulting in the production of high amounts of sulphide, coupled with high carbon dioxide values leading to drastic reduction in the incidence and abundance of plankton, benthic fauna and the fishery resources. Majority of these backwaters has a highly stressed environment, especially during the pre monsoon period when the retting activity is at its peak. The study has clearly reflected that a more restrained and cautious approach is needed to manage and preserve the unique backwater ecosystems of the country.

Introduction

The Convention on Wetlands of International importance is an intergovernmental treaty adopted in 1971 in Ramsar, Iran. It is known as the *Ramsar Convention* or the *Convention on Wetlands*. It provides a framework for international cooperation for the conservation and wise use (or sustainable use) of wetlands. As on 31st January 2005 the convention was signed by 144 Contracting Parties (nations) to the Convention, with 1420 wetland sites, totaling 123.9 million hectares were included in the Ramsar List of Wetlands of International Importance. Sustainable utilisation is understood as "human use of a wetland so that it may yield the greatest continuous benefit to present generations while maintaining its potential to meet the needs and aspirations of future generations". As on 31 January 2005, nineteen wetlands were designated under the Ramsar sites from India. In Kerala, the designated Ramsar sites are *Ashtamudi* (61,400 ha), *Sasthamcotta Lake* (373 ha) and *Vembanad-Kol Wetland*, Kerala (151,250 ha). Plate 1 gives the Ramsar sites of Kerala and other important wetlands. These waterbodies are have unique biological importance, forming a chain of the backwaters / estuaries / wetlands (locally called as *kayals*) along with the canals on the south-west coast of India. These interconnected backwaters together are a unique ecosystem supporting high biodiversity and a rich commercial fish and shellfish fauna. The available literature mainly pertains to the ecological characteristics, seldom taking into consideration the interrelationship with the fishery. This contribution reveals the environmental and fishery investigations in 11 backwaters in Kerala during 1996-98 periods on a seasonal basis (monsoon, pre monsoon and post monsoon) (Fig. 1). The backwaters investigated are given in next page.

	Backwaters	Abbr. Used	District	Area (ha)
1	Kadinamkulam	KDK	Trivandrum	347
2	Achuthengu	ANG	-do-	552
3	Ashtamudi	AST	Kollam	6424
4	Kayamkulam	KYK	Kollam/Alp	1652
5	Azhikode	AZK	Thrissur	696*
6	Chettuva	CTV	Thrissur	714
7	Ponnani	PNI	Thrissur	908**
8	Kadalundi-Beypore	KDL/BPR	Kozhikode	1192
9	Mahe	MHE	Kannur	88
10	Valapattanam	VPM	Kannur	3074
11	Neleswaram	NEL	Kasargod	825

* Kodungallur-Azhikode estuary; ** Ponnani + Puthuponnani

Source : Kerala Fisheries – An overview, 1992. Department of fisheries, Govt. of Kerala.

Materials and Methods

Water samples were collected and analysed for pH, dissolved oxygen, total sulphide, BOD_5, COD, free carbon dioxide, alkalinity, conductivity, total dissolved solids, nitrate, nitrite, phosphate and silicate (APHA, 1992). Rate of primary production was estimated using the dark and light bottle method as described by the Strickland and Parsons (1972). Plankton was collected using bolting silk having 56μ meshes. The phytoplankton, zooplankton and benthic fauna were collected and analysed based on Davis (1955), Ward and Whipple (1959), APHA (1992). The diversity index (H), dominance index (c), richness index (d) and evenness index (e) of plankton and benthos were also computed (Margalef, 1968). Fish fauna were identified based on Talwar and Jhingran (1991). Samples were collected from 33 stations at these kayal systems for the ecological assessment on a seasonal basis and 36 landing sites for the fishery survey. A landing centre approach was adopted for the fishery enumeration, and almost all the important landing sites around the lakes were covered for the survey (Gupta *et al.,* 1997).

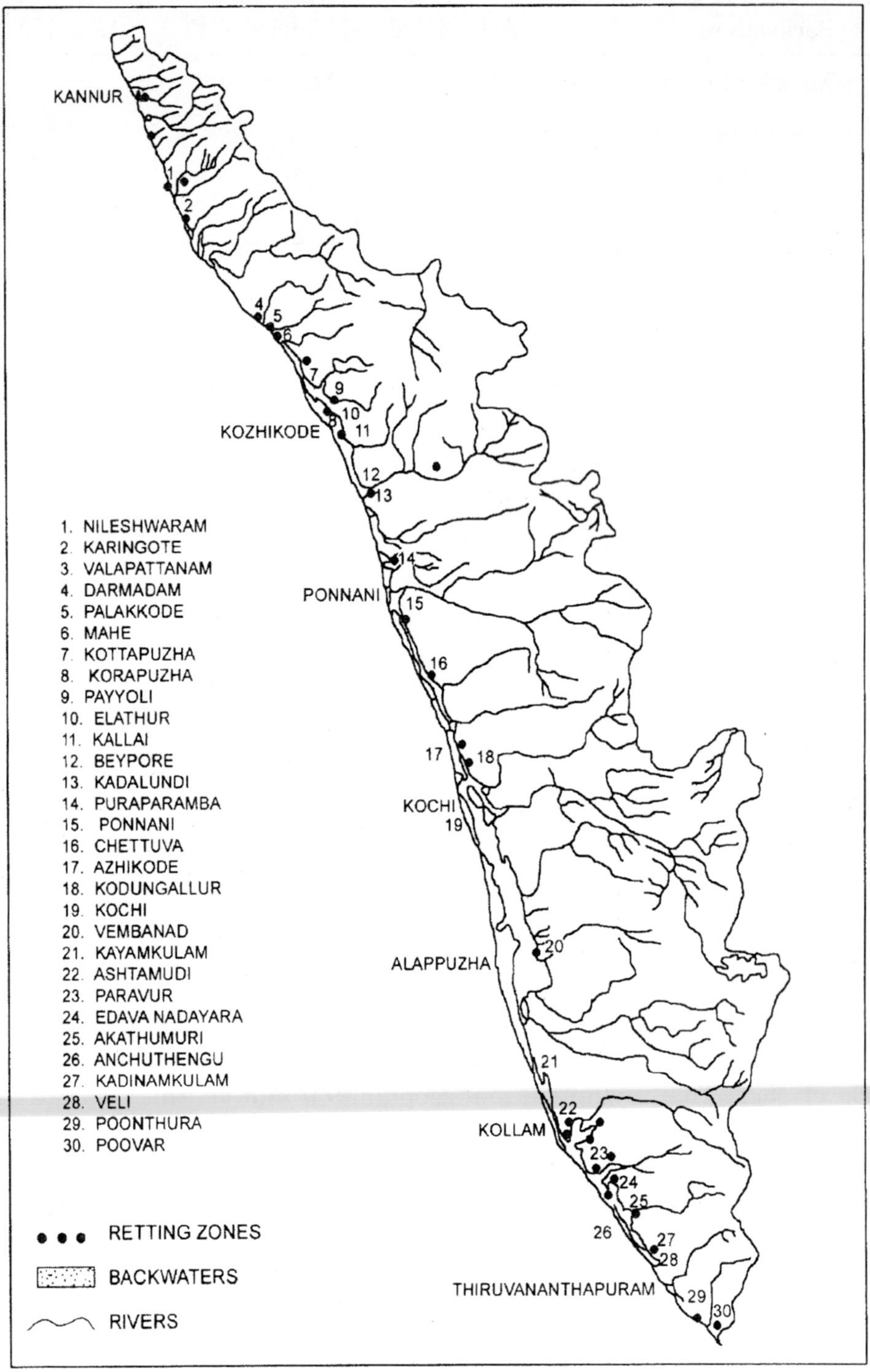

Fig. 1. Map of Kerala indicating the major backwaters, rivers & retting zones

Results and Discussion

Water Quality

The water quality parameters were recorded from the eleven backwaters during the pre-monsoon, monsoon and post monsoon periods (Figs. 2, 3 & 4).

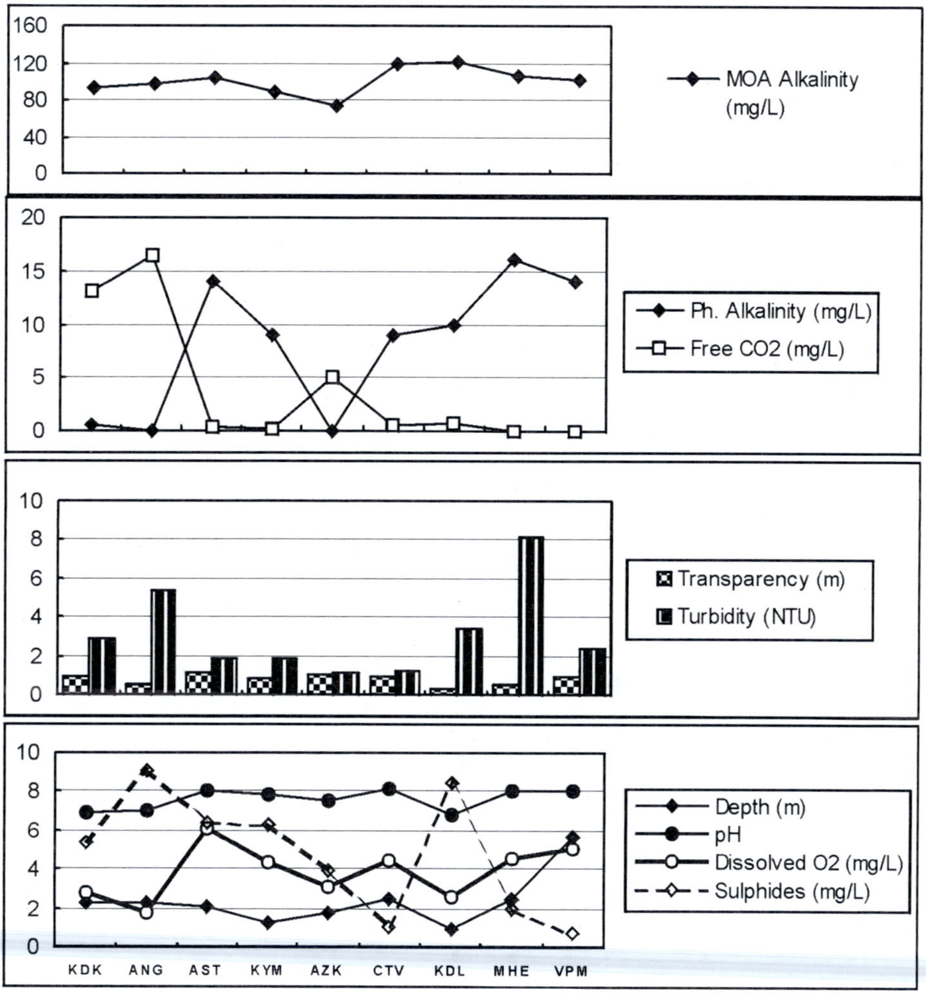

Fig. 2. Mean water quality variations in selected backwaters during pre monsoon period.

The pH was generally near neutral to alkaline in range. However, there was a reduction in its value particularly during the pre-monsoon period, owing to less mixing coupled with the impact or retting activity at certain stations. Anchuthengu, Kadinamkulam, Azhikode, Kadalundi and Chettuva recorded lower pH values. This was mainly due to the organic acids liberated during the retting. The mean pH values varied from 6.85 in Kadinamkulam to 8.12 in Chettuva backwater during pre-monsoon season. During monsoon, the variation was from 6.63 in Azhikode to 7.68 in Ashtamudi, whereas it varied from 6.74 in Ponnani to 8.20 in Neeleswaram during the post-monsoon period.

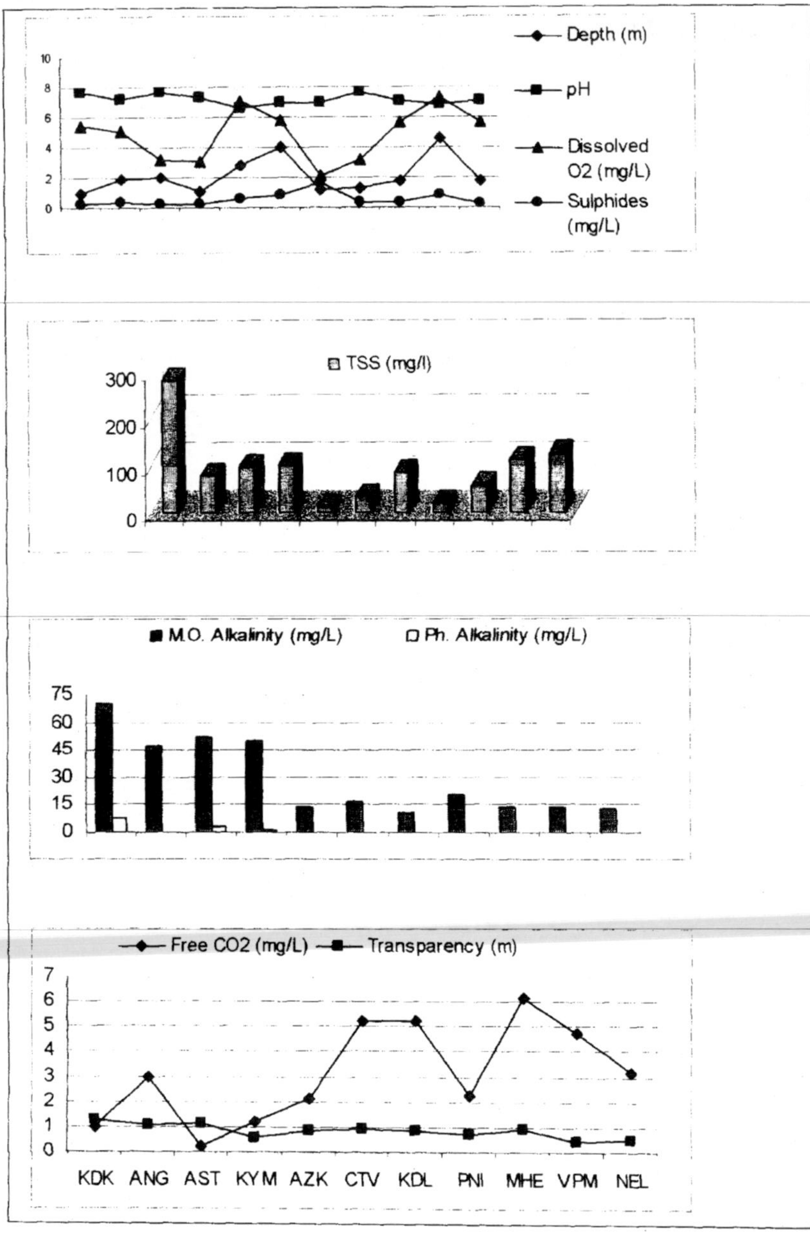

Fig. 3. Mean water quality variation in selected backwaters during the monsoon period

Moderate to low transparency values were observed in the systems (0.29-1.54 m). Retting areas had significantly lower values, particularly during the pre-monsoon period due to the accumulation of coir pith and ret liquor containing organic acids like pectin, pentosan, phenol, tannin, etc. in the water body. Turbidity from runoff substantially reduced the transparency in these shallow systems during monsoon while organic pollution resulted in low transparency during the pre-monsoon season. The salinity values indicated mixo-haline condition of these systems with a range of 5.20-32.38 ppt during the pre-monsoon, 0.18 to 22.42 ppt during the monsoon and 0.5 to 28.6 ppt during the post-monsoon periods. Such high variation was also dependent on the tidal impact and sampling time.

The dissolved oxygen level did not exhibit wide variation during the monsoon while nil to low values were observed at certain stations during the post-monsoon and the pre-monsoon periods. The highly stressed environment was evident from the fact that on an average, 7 out of the 11 investigated water bodies recorded dissolved oxygen within a range of 1.73-4.57 ppm during the pre-monsoon survey. Marked depletion of dissolved oxygen leading to anoxic condition coupled with the presence of sulfide was the most conspicuous observation at certain stations. This was mainly due to the intense ratting activity in these zones. The dissolved sulphides had an alarming concentration level in six of the systems during the pre-monsoon days. Sulfides exceeding 10 mg/L coupled with nil -<2.0 mg/L dissolved oxygen was prevalent in 12 of the 32 stations investigated during this season. Bottom layers of five sampling stations recorded no dissolved oxygen during the pre-monsoon months. This has been despite the exposure of the sampling stations to the tidal amplitudes twice a day.

The monsoon showers brought nutrients from allochthonous sources into the systems elevating the phosphate, nitrate and silicate concentrations in the water. Moderate to high COD values were observed during the post-monsoon when compared to the monsoon season. The high COD load could be due to the intense of retting activity and runoff from the surrounding areas.

The study indicated that retting of coconut husk in the back-waters has been the most contributing factor to the organic pollution leading to the depletion in the faunal resources in the backwaters as by the observations especially during the pre-monsoon survey. Nine of the eleven systems investigated were subjected to rampant retting activity.

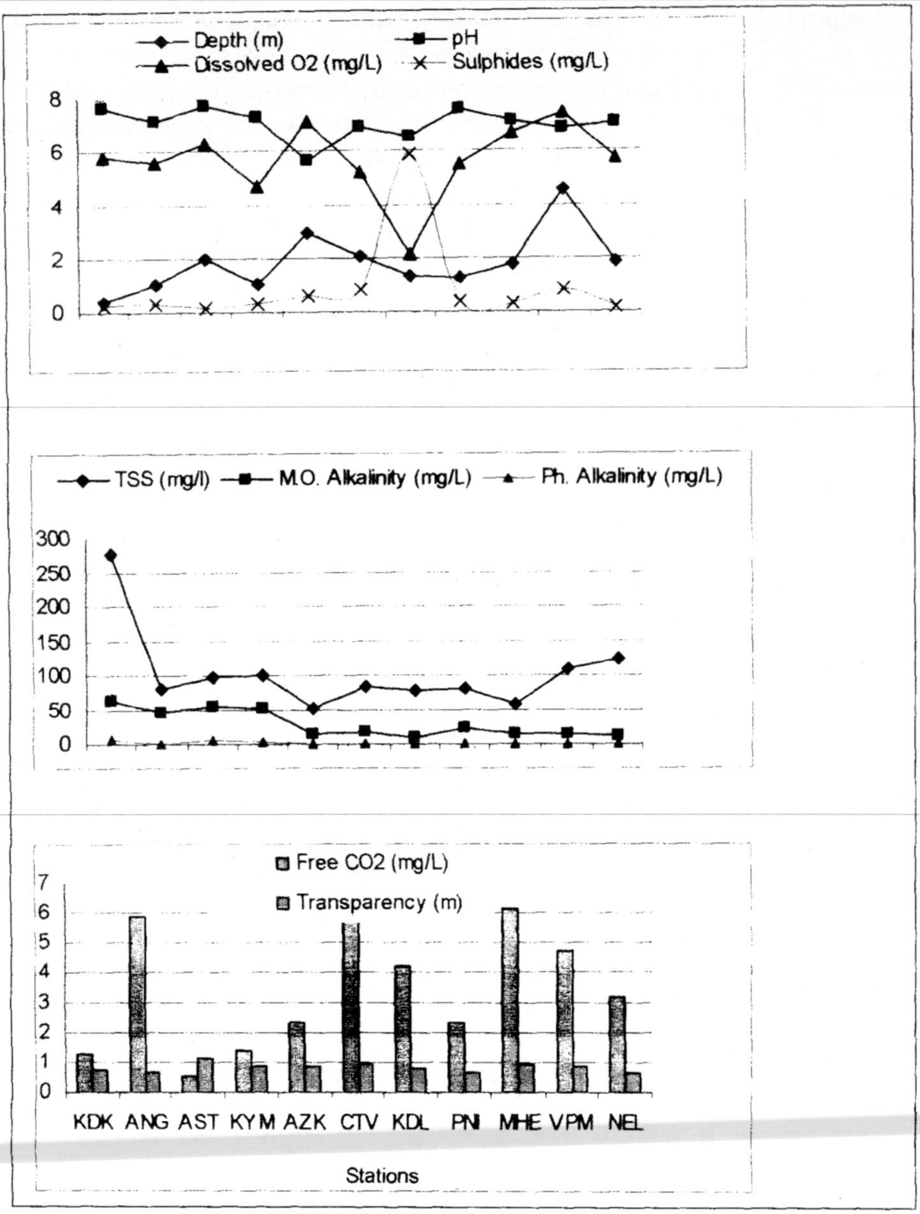

Fig. 4. Mean water quality variation in selected backwaters during the post monsoon period

The productivity values were generally low in most of the backwaters during the seasonal study. The gross production varied from a mean of 0.30 gC/m ^{3}/day in Anchuthengu to 0.98 gC/m ^{3}/day in Azhikode kayal. Earlier investigations conducted on the primary productivity in the Kadinamkulam backwater reported zero values at retting zones during October to May and 0.02-1.49 gC/m ^{3}/day at other stations and that in Ashtamudi the gross production rate was estimated at 143.88 mgC/m^3/hr (Bijoy Nandan, 2004). The present results were in conformity with these studies. Anchuthengu kayal had consistently poor chlorophyll values indicating the severe stress originating from coconut husk retting and other anthropogenic influences.

Sediment quality

The sediment parameters showed wide variations between the different backwaters during the pre monsoon, monsoon and post monsoon periods. The temperature was generally high during the pre-monsoon followed by the post-monsoon and the monsoon periods (Table 1). pH showed acidic trends in many of the backwaters particularly towards the southern, represented by the Kadinamkulam, Anchuthengu, Ashtamudi and Kayamkulam backwaters. A lowest mean value of 3.61 was recorded in Anchuthengu backwater during the pre-monsson period. The acidic pH of the sediment could be due to the intense pectinolytic activity during retting of coconut husk in the backwaters resulting in lowering of the values. Conductivity and organic carbon values were also high during the pre-monsson period as compared to the other two seasons, indicative of the accumulation of organic matter and other materials during the period. The intense retting activity resulting in organic enrichment was also responsible for the higher organic content of the sediment in the southern backwaters as compared to the northern. Calcium carbonate was at its peak in Ashtamudi (Av. 17.65%) during the pre-monsoon and the lowest in Anchuthengu backwater (Av. 0.29%) during the monsoon period. The mining activity of clams and oysters was more prevelant in the southern backwaters which could have resulted in higher values of calcium carbonate in the sediment. The available phosphorous content was high during the pre-monsoon with an average of 0.40% during the pre-monsoon followed by monsoon (Av. 0.27%) and post-monsoon (Av. 0.28%) in the ten backwaters.

Sand fraction dominated in all the backwaters investigated followed by clay and silt. The northern backwaters were higher in the sand content as compared to the southern systems (Table 1).

The post-monsoon period was characterized by the peak sand fraction followed by pre-monsoon and the monsoon periods. Coarse sand also formed an important constitutent of the sediment texture contributing a mean of 50.80% (Av.) during the monsoon, 38.49% (Av.) during the post-monsoon and 45.91% (Av.) during the pre-monsoon. Silt distribution in the backwaters were very uneven without any steady pattern, where Ashtamudi had the highest average in the all the three seasons followed by Kadalundi during the pre-monsoon and Kayamkulm during the monsoon and post monsoon periods. The clay fraction was maximum in the southern backwaters during all the three seasons with expectionally high mean value in Anchuthengu (Av. 39.15%). This could be due to the high organic enrichment due to retting activity as well as the stressed condition due to the dry pre-monsoon months prevelant in the backwater.

Table 1. Mean percentage variation in sediment texture in selected backwaters during 1997-99 period

PRE-MONSOON											
	NEL*	VPM	MHE	KDL	CTV	AZK	KYM	AST	ANG	KDK	MEAN
Sand	0.00	33.94	34.20	37.12	31.22	37.26	53.42	30.47	28.49	38.26	36.04
Coarse sand	0.00	50.44	50.10	20.61	52.38	50.63	25.72	16.16	28.24	28.92	35.91
Silt	0.00	2.75	10.10	11.80	3.67	2.67	3.00	16.50	2.75	9.08	6.92
Clay	0.00	12.32	17.70	29.78	12.23	8.90	16.98	37.70	39.15	20.48	21.69
	* sampling not undertakern										
MONSOON											
	NEL	VPM	MHE	PNI	CTV	AZK	KYM	AST	ANG	KDK	MEAN
Sand	75.36	36.38	40.21	29.93	19.81	51.83	10.27	22.74	48.46	21.15	35.61
Coarse sand	19.39	54.27	52.25	67.50	55.35	37.47	59.53	45.71	45.98	70.59	50.80
Silt	2.18	3.55	2.08	1.25	5.40	1.45	4.75	16.83	2.77	3.50	4.38
Clay	2.57	4.90	5.07	1.21	11.96	8.45	21.34	13.86	2.30	10.59	8.23
POST-MONSOON											
	NEL	VPM	MHE	PNI	CTV	AZK	KYM	AST	ANG	KDK	MEAN
Sand	53.62	45.40	46.41	50.76	25.45	47.57	45.48	7.45	48.60	27.52	39.83
Coarse sand	27.41	37.91	31.02	38.39	47.05	44.14	19.29	49.84	48.60	41.23	38.49
Silt	8.10	3.82	2.83	1.31	6.56	1.55	10.81	16.71	3.81	5.29	6.08
Clay	10.44	12.17	9.57	8.78	20.71	6.05	23.90	25.19	13.00	19.09	14.89

Impact of retting activity in the wetlands

There are several anthropogenic interventions affecting the sustainability of fisheries development in the country. Some of them are effluent discharge from factories/industries, organic pollution from various sources and a host of others. But, retting of coconut husk for the production of coir is the most extensive and the most singular form of pollution on the south west coast of the country, affecting the entire backwater ecosystems of the region. Retting, is basically a "*soaking process*" where husks are arranged in bundles in huge coir nets known as "*malis*" and allowed to float freely in the backwaters, until they get soaked, become heavy and gradually sink to the bottom. Retting is a biological process involving bacteria, fungi and yeast. Large chunks of pectin, phenol, cellulose, hemicellulose and tannin are released from the husk into the surrounding medium, during different stages of retting. In depth studies on the impact of retting activity on the backwater systems, was reported by Bijoy Nandan (2004).

Acidic pH condition coupled with anoxia and production of high concentrations of hydrogen sulphide were the outstanding feature, in the environmental quality of the retting zones. The higher concentration of free CO_2 in the retting zones could be attributed to the process of decomposition of organic matter like pectin, phenol, tannin etc. leading to a rise in temperature of the medium, thereby favouring production of the gas (Table 2).

Table 2. Mean values of physico-chemical characteristics of retting and non-retting zones in the backwaters

	Retting zone	Non-retting zone
Depth (m)	1.88	2.88
Transparency (m)	0.60	0.69
pH	6.92	7.99
Dissolved oxygen (mg/L)	2.43	7.60
Total sulfides (mg/L)	8.80	3.01
Turbidity (NTU)	2.62	1.60
Free carbon dioxide (mg/L)	6.4	3.5
Alkalinity (MO) (Mg/L $CaCO_3$)	103	91
Alkalinity (Ph.) (Mg/L $CaCO_3$)	6.7	10
Inorganic phosphate (mg/L P)	40.6	36.5

The primary productivity mechanism was totally collapsed in the retting zones. It has thus shown that the productivity potency of the coastal ecosystems was adversely affected due to pollution from retting activity. The mean chlorophyll a (1.63 mg/m^3) and algal biomass (1.09 g/m^3 wet wt.) values were very low in the retting zones as compared to the non-retting zones (chlorophyll a: 9.65 mg/m^3; algal biomass: 6.46 g/m^3 wet wt.). Studies by Bijoy Nandan (2004) have shown that the plankton, benthic fauna and fish biodiversity showed massive depletion in the retting zones as compared to the non retting zones.

Mass mortality of fish and shell fishes were reported from the retting zones, particularly during the pre monsoon period affecting the sustainable fish production in this region as well as the adjacent areas.

Biodiversity and Community Structure

The biomass values of plankton in the retting zones, were greatly reduced, the lowest in the retting zones of Kadinamkulam (0.4 ml/L) and the highest in the non- retting zones of Valapattanam (12.8 ml/L). The incidence, abundance and diversity of fauna were greatly depleted in the retting zones as compared to the non-retting zones. This depletion was more prominent during the pre monsoon period when the retting process attained its peak, resulting in anoxic conditions coupled with the formation of high concentrations of sulphide in the medium. The diversity index (H), richness index (d) and evenness index were generally low in the retting zones.

The mean seasonal biomass values varied from 0.52 in Azhikode to 5.85 mL/m^3 in the Kadinamkulam backwater during the monsoon, whereas it varied from 0.28 in Mahe to 6.83 mL/m^3 in Ashtamudi estuary during the post-monsoon period. The high amount of detritus, sediment and other suspended materials collected along with the plankton samples during the monsoon showers resulted in higher settling volume in this period. This could be the reason for the higher biomass recorded during the monsoon period. The summer values wer characterised by higher biomass volume in general, but a reduced volume at retting areas. Thus the highest value was obtained at Valapattanam (12.8 ml/m^3) and the lowest at Kadinamkulam (0.4 mL/m^3).

Phytoplankton and Zooplankton

The Kadinamkulam estuary showed the maximum mean phytoplankton population during the monsoon period whereas the Chettuva estuary showed the maximum value in the post-monsoon period (Table 3).

Desmidaceae had a higher representation in the northern backwaters (Neeleswaram to Azhikode) during the monsoon season whereas this was replaced by either Bacillariophyceae or Chlorophyceae during the post monsoon season. The southern backwaters except Ashtamudi were dominated by Bacillariophyceae during monsoon, which got replaced by Myxophyceae during the post monsoon season. During the pre-monsoon period Chlorophyceae, Myxophyceae and Chrysophyceae showed higher percentage incidence in the southern backwaters whereas Bacillariophyceae showed higher incidence in the northern backwaters. *Campylodiscus* sp., *Stauras-*

trum sp., *Micrasterias* sp., and *Spondylosium* sp. contributed to the higher density of Bacillariophyceae. Chlorophyceae represented by *Microspora* sp., *Pediastrum* sp. and *Hormidium* sp., contributed to the high planktonic biomass in Chettuva and Ponnani estuaries.

Table 3. Percentage distribution of Phytoplankton in selected backwaters during the seasonal survey

Backwaters	NEL	VPM	MHE	KDL	CTV	AZK	KYM	AST	ANG	KDK
				PRE-MONSOON 1996						
Chlorophyceae		-	-	1.6	-	3.3	-	0.5	13.3	-
Myxophyceae		-	1.1	6.4	-	3.3	30.4	62.6	20.0	82.1
Chrysophyceae		6.9	-	31.4	3.5	-	56.5	19.0	-	1.5
Desmidaceae		-	-	-	-	-	-	-	20.0	-
Bacillariophyceae		93.1	98.9	60.6	96.5	93.4	13.1	17.9	46.7	14.9
Pyrrhophyceae		-	-	-	-	-	-	-	-	1.5
TOTAL (No/m³)		27250	17750	6218	91700	6600	2300	12930	1125	7763
Backwaters	NEL	VPM	MHE	PNI	CTV	AZK	KYM	AST	ANG	KDK
				MONSOON-1997						
Chlorophyceae	39.8	11.1	14.0	44.2	35.9	26.9	3.6	2.7	10.7	-
Myxophyceae	0.8	2.9	2.5	23.5	10.3	5.3	2.3	90.0	2.6	1.3
Chrysophyceae	-	-	-	0.4	0.6	-	-	0.7	-	-
Desmidaceae	43.5	66.3	62.3	27.3	36.1	58.8	6.2	0.6	29.0	2.5
Bacillariophyceae	16.0	19.7	21.2	4.7	17.0	9.0	87.9	4.5	57.8	93.2
Pyrrhophyceae	-	-	-	-	-	-	-	1.5	-	3.0
TOTAL (No/m³)	13748	26078	24879	6099	4823	2895	23025	57780	20933	71363
				POST-MONSOON 1997						
Chlorophyceae	4.9	71.7	1.9	29.1	92.7	22.5	11.3	0.6	13.8	4.3
Myxophyceae	0.8	9.4	3.8	16.7	1.8	9.3	32.7	32.5	47.4	45.9
Chrysophyceae	-	-	-	1.8	0.1	-	4.0	-	0.8	-
Desmidaceae	-	-	2.9	47.7	3.8	19.4	16.7	0.4	23.0	2.8
Bacillariophyceae	85.6	18.9	88.4	4.7	1.6	47.1	33.1	5.5	15.1	46.9
Pyrrhophyceae	8.7	-	2.9	-	-	1.8	2.2	60.9	-	-
TOTAL (No/m³)	6600	795	3901	65438	81188	6810	10314	79375	23164	37075

A total of 100 species of phytoplankton were recorded from the backwaters. The diversity index (H), richness index (d) and evenness index (e) of the phytoplankton showed peak values during the pre monsoon and post monsoon periods in the Valapattanam and Azhikode backwaters whereas minimum values were observed in the Kadalundi, Chettuva and Ashtamudi backwaters. The monsoon period showed the presence of 14 groups of zooplankton whereas the post monsoon showed 20 groups in the backwaters. But during the pre-monsoon study conducted in the same water bodies during 1996 showed the presence of only 12 faunal groups. The southern back-waters (Kayamkulam to Kadinamkulam) showed higher incidence and diversity of the different planktonic groups when compared to the backwaters in the nortern segment during monsoon as well as post-monsoon periods. During the pre-monsoon, the Kayamkulam backwater recorded the maximum numerical density (64273 Nos./ m^3) and the Mahe backwater recorded the minimum (3350 Nos./ m^3) (Fig. 5). *Centropyxis* sp., the protozoan showed its maximum incidence in the Mahe and Azhikode estuaries during the monsoon period. The retting zones in the ten ecosystems showed considerably lower planktonic abundance and diversity, coinciding with the poor water quality condition in the corresponding areas. Azhikode recorded the highest diversity index value (H) during the monsoon period (2.74) and the lowest in Kayamkulam during the same season. Copepods and copepod nauplii formed an important component in all the ten systems in both the seasons. In the Neeleswaram backwater, 50% of the plankton were contributed by copepods during the post-monsoon period whereas the group formed 34.8% of the population during the monsoon period in the Kadinamkulam estuary. Altogether, 34 species of rotifers were recorded during the post-monsoon season alone in the ten backwaters. *Brachionus* species represented by B.*plicatilis, B. falcatus, B. calyciflorus* showed the maximum incidence among the rotifers in the present study.

Benthic fauna

Amphipoda, Polychaeta and Gastropoda formed the dominant groups in all the backwates during both the seasons. The monsoon as well as post monsoon periods showed higher numerical density in the southern backwaters when compared to the northern segments. Nemertea (Ribbon worms), a rare group was recorded in the Neeles-waram (0.5%) and Ashtamudi backwaters (0.4%) during the post-monsoon period (Fig. 6). The pre-monsoon was charecterised by higher groups of benthic fauna like Oligochaeta, Polychaeta, Amphipoda, Insecta and Mollusca in the northern backwaters when compared to the backwaters in the southern side. The post monsoon period showed higher diversity index (H) and richness index (d) values when compared to the pre-monsoon and monsoon periods. As observed above in zooplankton, the benthic fauna also showed considerable depletion in the retting zones of the present study. Thirteen groups formed the benthic population during the monsoon period whereas seventeen groups in the post-monsoon. Insect fauna showed higher incidence and diversity particularly in the backwaters in the northern side during the post-monsoon period. The retting zones in the backwaters showed higher dominance of *Chironomus* larvae and other insects.

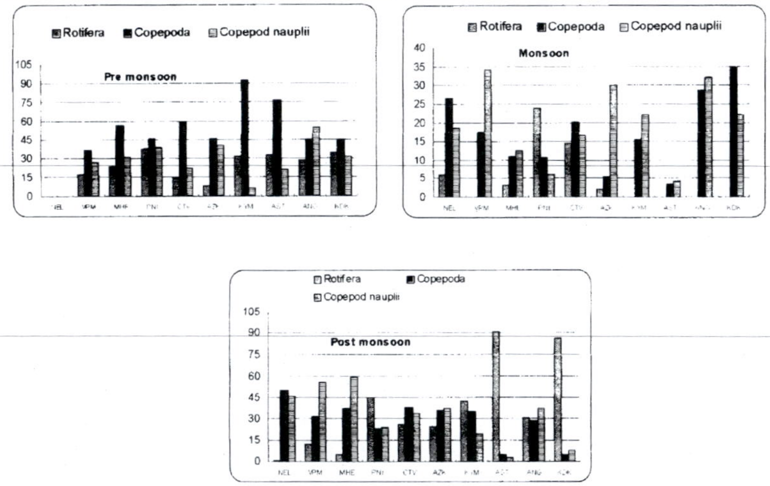

Fig. 5. Seasonal mean distribution (%) of selected zooplankton in the backwaters

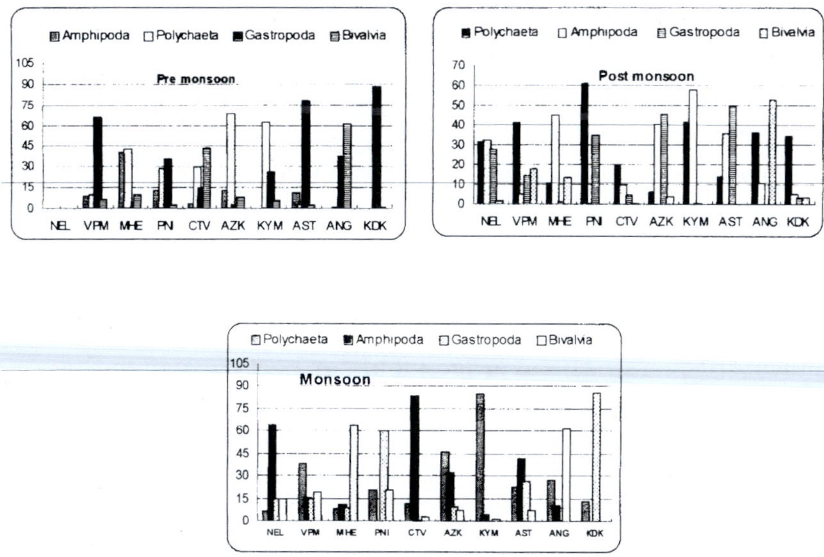

Fig. 6. Seasonal mean distribution (%) of selected benthic fauna in the backwaters

The benthic forms were present at all seasons showing its maximum mean density (39 nos./m^2) during the monsoon. The high organic carbon content in combination with the higher percentage of sandy clay during the post-monsoon and pre-monsoon periods was a positive element in maintaining a high density of the benthic organisms. The proboscis worms or the nemertineans belonging to the phylum Nimertinea was one of the rare groups of benthic fauna recorded during the present investigation. This is the first record of this worm from the inland waters of this geographic region.

Fish Production Potential

Ninety four species of fish and shellfish were identified that contributed to the fishery of these kayals. Definite zone marking could be seen in the distribution of these species in the backwater. *Puntius filamentosus, P. sarana, Labeo dussumieri, Mystus malabaricus, Anabas testudineus, Channa* spp., *Oreochromis mossambicus* and *Mastacebelus armatus* could be cited as examples. *Rhinobatus halavi, Congressox talabanoides, Lobotes surinamensis, Acanthurus strigosus, Eleotris fusca, Lepturocanthus savala, Platax orbicularis*, etc were also recorded from certain systems, though in stray numbers. *Etroplus suratensis, Penaeus indicus* and the *Metapenaeus monoceros* represented 2.0-13.9%, 2.5-29.6% and 1.0-8.2% in the total catches (Table 4). The total landings from different backwaters varied from 96.8t from Mahe to 2899t from the Ashtamudi. The average yield/ha varied from 410 kg at Anchuthengu to 2747.3 t from Azhikode estuary. The CPUE exhibited wide variation from gear to gear, but exhibited a general trend for most of the backwaters. The high rate CPUE for the Stake net (7.5-18.1) was largely offset by the limited days of operation and the low market value fish forming the bulk of the catch. Seine net (11.3-74.7) too brought high CPUE, but again the catch composition was composed largely of small fishes, young fishes and the low priced *M. dobsonii.*

Therefore the investigation clearly indicated that from the 15,100 ha backwater covered, the average yield per hectare was to the tune of 651 kg. Considering that the coastal interconnected backwaters spread to over 65,000 ha, the total annual yield should exceed 42315 tones. The average density of active fishermen being 74 per km^2, the total fishermen directly depending on the backwater fishery is estimated at about 50,000 nos. Therefore the fishermen population directly depending on the backwater fishery approximates 2.5 lakhs. annual yield should exceed 42,315 tones.

Table 4. Percentage contribution by various species/groups to the total landings from selected kayals

Species/ Groups	Contribution (%) to total landings	
	Range	Mean% of Pooled data
Acanthurus spp.	0.0-4.9	0.47
Gerres spp.	1.0-6.1	2.72
Platycephalus sp.	0.1-3.1	1.08
Leiognathus spp.	0.3-5.9	1.08
Etroplus spp.	2.0-13.9	5.68
Megalops sp.	0.0-5.8	0.60
Tachysurus spp.	1.9-10.4	3.20
Ambassis sp.	0.0-8.6	2.92
O. mossambicus	0.0-14.7	0.94
Stolephorus sp.	0.0-3.3	1.57
Sillago sihama	0.0-3.8	0.81
Caranx spp.	1.0-5.2	1.29
Lutjanus spp.	1.1-11.4	1.50
Mullets	1.5-16.5	5.07
Flat fishes	0.1-3.6	1.21
Half beaks	0.1-2.5	0.26
Others	6.4-20.0	10.66
Fishes total	26.9-75.0	41.20
Metapenaeus dobsonii	9.2-53.4	33.06
M. monoceros	1.0-8.2	6.53
Penaeus indicus	2.5-29.6	9.19
P. monodon	0.0-9.0	1.22
Other penaeids	0.0-2.9	0.64
Non-penaeids	0.0-1.3	0.46
Prawns total	13.9-70.5	53.1
Crabs	2.6-11.1	5.75
Total yield (kg/ha.)	246-2747	630.1

Major Recommendations from the Study

1. The fishing effort should not be allowed increase further and has to be restricted at least to the current level till further

suggestions are made based on population dynamics investigations conducted on major fish/shellfish species of the backwater systems.

2. No proper registration of the fishing gear and craft is being carried out except for the stake nets. The registration system followed under the Fishermen Welfare Board is not effective in regulating the fishery. Strict registration and licensing to all existing craft, gear and fishermen is to be immediately implemented.
3. There is an urgent need to restrict the mesh size of the stake net, Chinese dip net and the drag net to ensure more growing period to the young ones. Though a minimum mesh size of 25 mm (stretched) is advisable, considering that *M. dobsonii* is also to be exploited, the minimum mesh size may be restricted to 18 mm.
4. It has been observed that several units of purse seine are diverted to the backwaters during the closed season for marine fishing. This has to be totally prohibited.
5. Several of the stake nets are being deployed during the tide incursion to the back-waters against the norms. The enforcement machinery is to be strengthened to ensure that the stake nets are deployed only during the receding phase.
6. Considerable area of the backwaters has already been lost due to reclamation for agricultural, mining, urban area development and similar activities. Backwaters are priceless heritage serving to a variety of economic activities apart from fisheries. Further encroachment/reclamation are to be strictly regulated.
7. Several stretches of backwaters are subjected to extreme organic/industrial pollution. Hence pollution abatement measures are to be given top priority. Technology for alternate coconut husk retting practice has to be developed that free backwaters from organic pollution.
8. Reclaimed paddy lands such as at Kuttanad, Kattampally, etc are to be utilised to raise an additional crop of fish during the fallow period.
9. Ranching programme will be more effective after the implementing fishery regulations. Hence regulatory measures are to be strictly implemented to benefit from the currently envisaged massive ranching programme in backwaters/river stretches.

Acknowledgement

The author is thankful to the Director, Central Inland Fisheries Research Institute (CIFRI) for providing the facilities and also to the staff for giving unstinted support in undertaking the work.

References

Apha, 1992. American Public Health Association Standard Methods for the examination of analysis of water and waste-water. 19th edition.

Bijoy Nandan, S. 2004. *Studies on the Impact of retting on Aquatic Ecosystems*, ISBN 81-901939-0-2, Limnological Association of Kerala, India, 120 p.

Davis, C.C., 1995. The Marine and Fresh water plankton, Michigan State University Press.

Gupta, R.A., S.K. Mandal and S. Paul 1997. *Methods of collection of Inland Fisheries Statistics in India,* Part-1, Survey Methodology Guidelines Bulletin No. 77, CIFRI Publication. p. 64.

Margalef, R. 1968. *Perspectives in ecological Theory*. University of Chicago press, Chicago.

Strickland, J.D.H and T.R. Parsons 1972. *A practical handbook of sea water analysis*, Bulletin No. 167, Fisheries Research Board of Canada.

Talwar, P.K. and A.G. Jhingran 1991. *Inland Fishes*, Vol. 1 & 2 Oxford & IBH Publishing Co. Pvt. Ltd.

Ward, H.B and G.C. Whipple, 1959. Freshwater Biology. Second Edition, John Wiley and sons, Inc. p. 1248.

THE RED LISTED FLOWERING PLANTS IN KERALA AND THEIR CONSERVATION STATUS

N. Sasidharan

Kerala Forest Research Institute, Peechi - 680 563, Thrissur, Kerala

The Kerala State lies along the southwest corner of Peninsular India has an area of 38863 km^2, which is about 1.18 per cent of the total area of the country. The State has a complex topography with mountains, valleys, ridges and scarps. The altitude varies from sea level to 2695 m above msl – Anamudi, the highest peak in Peninsular India. The State receives both southwest and northeast monsoons. The mean annual rainfall is 2693 mm (Kerala Land Use Board, 1997). The varied topographical features, high rainfall and geologic conditions favoured the formation of different ecosystems from shola forests on the mountains valleys to the mangrove forests along the seacoasts and estuaries. The most outstanding feature of the State is the formation of tropical rain forests along the windward side of the Western Ghats, which is lying parallel to the west coast. A small extent of the area is along the rain shadow region of the Western Ghats where the vegetation is dominated by dry deciduous forests and scrub jungles. The State has a sizable area of wetlands consisting of ponds, lakes, lagoons and marshy areas. Thus, the State is endowed with a very diverse and rich flora, which account for nearly 25 per cent of the flora of India. Angiosperms are the dominant plant group and they are reasonably well explored.

The floristic studies of most of the districts and Protected Areas (PAs) in Kerala have been carried out during the last three decades. Besides, these general floristic studies, revisionary studies of several genera and a few families were also carried out. These studies have brought out several new taxa as well as new record of occurrence. Manilal and Raveendrakumar (1998) have compiled a list of 625 taxa of flowering plants as additions to the *Flora of the Presidency of Madras* (Gamble and Fischer, 1915-1936). A comprehensive checklist of the flowering plants of Kerala (Sasidharan, 2004) reveals the occurrence of 4679 taxa including 458 exotics naturalised or grown as Plantation/Agriculture/Horticultural crops in Kerala.

The Western Ghats is considered as a biodiversity hotspot of endemism. The forests of Kerala are lying along the southern Western Ghats, which is considered as the richest region with respect to species richness and endemism. Nayar (1996) identified three hotspots of endemic centres in Kerala, viz., Agasthyamala, Anamala-High Ranges and Silent Valley-Wyanad. Kerala is one of the densely populated States in India and therefore the biodiversity is under severe anthropogenic pressure. The State has lost considerable forest area due to expansion of agriculture/plantation crops, construction of river valley projects and other developmental activities. A study on the forest loss for Kerala indicates that between 1961 and 1998, 8.4 per cent of the forests were lost (Prasad, 1998). The forest cover loss in Agasthyamala, one of the hotspots of endemic centres during 1960-1990 is 9.9 per cent (Ramesh *et al.*, 1997). The impact of deforestation is increasingly felt in recent years. Landslide is common during monsoon. The degradation of natural habitats has adversely affected the population of endemic taxa, which are narrowly distributed. According to Ramesh *et al.* (2003) nearly 51 per cent of the forests of Kerala are degraded. Another alarming factor is the lack of regeneration in the forests particularly in the moist deciduous forests. Forest fires invariably caused by man, frequent during summer often wipe out the entire ground flora consisting of herbs, shrubs and seedlings of trees. The forest fire further promotes the growth of light demanding aggressive exotic weeds. These weeds replace the indigenous characteristic species of the forests. The most aggressive weeds in the forests are *Chromolaena odorata, Mikania micrantha, Lantana camara, Ageratina adenophora, Hyptis suaveolens, Mimosa invisa,* etc. *Parthenium hysterophorus* is mostly established in the forests of the low rainfall areas. Weed growth is very severe in forest plantations. All these factors have contributed to the rarity of species.

IUCN has estimated that about 10 per cent of the vascular flora is under threat and most of them inhabit in the tropics. The rare and threatened plants in the South India were assessed by Joseph (1977), Henry *et al.* (1979), Jain and Sastry (1984), Ahmedullah and Nayar (1987), Nayar and Sastry (1987, 1988, 1990), IUCN (2000) and Sasidharan (2003). Nayar (1997) listed 1272 Western Ghats endemic taxa in Kerala and 483 among them were placed under various threatened categories.

Based on the various assessments of the threatened flowering plants, as many as 492 species recorded from Kerala have been evaluated to belonging to threatened categories. As per the IUCN Red List categories 35 species are considered as Extinct in the wild - EX (Possibly Extinct); 137 Critically Endangered - CR; 147 Vulnerable - VU; 142 Low Risk/ - LR and 31 as Data Deficient – DD [Table 1]. Among the families Rubiaceae has the highest number of Red Listed species (67); Leguminosae (33); Balsaminaceae (32); Asteraceae (27); Orchidaceae (24); Asclepiadaceae (21); Poaceae (19); Myrtaceae (19: Araceae (18) and Melastomataceae (16) [Table 2]. *Impatiens* has the highest number of Red Listed species (32 species) followed by *Vernonia* (14) and *Hedyotis* (13) [Table 3].

Table 1. ICUN Red Listed Species Categories

Category	No. of species
Possibly Extinct (in the wild)	35
Critically Endangered	137
Vulnerable	147
Low Risk	142
Data Deficient	31
Total	**492**

Table 2. Family wise analysis of Red Listed species in Kerala

Family	Total no. of species	RET species
Rubiaceae	243	67
Leguminosae	381	33
Balsaminaceae	80	32
Asteraceae	182	27

Orchidaceae	261	24
Asclepiadaceae	72	21
Poaceae	409	19
Myrtaceae	59	19
Araceae	62	18
Melastomataceae	65	16
Euphorbiaceae	181	16
Lauraceae	64	14
Lamiaceae	116	10
Annonaceae	51	10
Commelinaceae	55	9
Anacardiaceae	26	9
Symplocaceae	19	8
Gesneriaceae	15	7

Table 3. Genus wise analysis of Red Listed Species in Kerala

Genus	Total no. of species	RET species
Impatiens	79	32
Vernonia	31	14
Hedyotis	38	13
Syzygium	36	12
Lasianthus	14	11
Ceropegia	20	11
Sonerila	23	10
Crotalaria	52	9
Pavetta	18	8
Ophiorrhiza	19	8
Arisaema	17	8
Symplocos	19	7
Exacum	14	6
Eugenia	10	6
Didymocarpus	9	6
Bulbophyllum	16	6
Begonia	11	6
Phyllanthus	23	6

Most of the critically endangered as well as extinct in the wild category species are known only by their type collections. Recent floristic studies have relocated a large number of threatened species. Despite the repeated efforts species such as *Bulbophyllum albidum, Burmannia indica, Burmannia stricta, Caralluma pauciflora, Caralluma procumbens, Christisonia keralensis, Dalbergia travancorica, Didymocarpus macrostachya, Didymocarpus meeboldii, Diospyros sulcata, Fergusonia tetracocca, Habenaria flabelliformis, Habenaria richardiana, Hedyotis beddomei, Hedyotis sisparense, Impatiens aliciae, Impatiens cochinica, Impatiens concinna, Impatiens orchioides, Impatiens pallidiflora, Impatiens platyadena, Litsea nigrescens, Ophiorrhiza barnesii, Ophiorrhiza incarnata, Ophiorrhiza pykarensis, Pavetta nemoralis, Pavetta travancorica, Phyllanthus megacarpa, Psychotria beddomei, Senecio multiceps, Sonerila wynaadensis, Vernonia fysonii,* etc. (list not exhaustive) are yet to be relocated.

The State of Kerala has a relatively high per cent of PAs. About 25 per cent of the forest area has been declared as PAs by establishing 12 Wildlife Sanctuaries, two National Parks and one Bird Sanctuary. An analysis of the flora of the PAs shows that 73 per cent of the Red Listed species occur in the Protected Areas. Among the Protected Areas, Periyar Tiger Reserve holds 151 Red Listed species, followed by Neyyar Wildlife Sanctuary with 142 species, Shenduruny Wildlife Sanctuary 101 species; Silent Valley 106 species and Parambikulam Wildlife Sanctuary with 109 species [Table 4].

Table 4. Sanctuary wise analysis of Red Listed species in Kerala

Sanctuaries/ National Parks	Area (km^2)	Total no. of species	Endemic	RET species
Periyar	777	1985	515	151
Neyyar	128	1080	384	142
Parambikulam	272	1436	360	109
Silent Valley	90	1104	392	106
Shenduruny	100	951	310	101
Peechi-Vazhani	125	829	164	34
Eravikulam	97	326	129	33
Chinnar	90	961	170	26

Apart from the PAs, the Sacred Groves outside the forest areas abode several endemics as well as Red Listed species. Induchoodan

(1998) enumerated 720 species of flowering plants from the 364 sacred groves of Kerala of which 154 are Western Ghats endemics. The Red Listed species recorded from Sacred Groves are *Aporosa bourdillonii, Begonia canarensis, Blepharistemma serratum, Cleome burmannii, Dysoxylum malabaricum, Goniothalamus wynaadensis, Kingiodendron pinnatum, Kunstleria keralensis, Myristica fatua* var. *magnifica, Myristica malabarica, Nothopegia aureo-fulva, Ochreinauclea missionis, Pterospermum reticulatum, Strophanthus wightianus, Syzygium chavaran, Syzygium mundagam, Syzygium occidentale, Syzygium travancoricum, Vateria indica,* etc.

References

Ahmedullah, M. and M.P. Nayar, 1987. Endemic Plants of the Indian Region. Botanical Survey of India, Calcutta.

Gamble, J.S. and C.E.C. Fischer, 1915-1936. The flora of the Presidency of Madras Adlard & Son Ltd., London.

Henry, A.N., K. Vivekanandan and N.C. Nair, 1979. Rare and threatened flowering plants of South India. J. Bombay Nat. Hist. Soc. 75: 684-697.

Induchoodan, N.C. 1998. Ecological Studies of the Sacred Groves of Kerala. Ph.D. Thesis, Central University, Pondicherry.

IUCN, 2000. The 2000 IUCN Red List of Threatened Species. IUCN, Gland.

Jain, S.K. and A.R.K. Sastry, 1984. The Indian Plant Red Data Book. Botanical Survey of India, Calcutta.

Joseph, J., 1977. Floristic studies in India with special reference to Southern Circle of Botanical Survey of India. Bull. Bot. Surv. India 19: 109-111.

Kerala State Land Use Board, 1997. Kerala State Resource Based Perspective Plan 2020AD. Kerala Land Use Board, Trivandrum.

Manilal, K.S. and K. Raveendrakumar, 1998. Addition to the Flora of Kerala since Gamble (1935). Rheedea 8: 179-241.

Nayar, M.P. and A.R.K. Sastry, 1987, 1988, 1990. Red Data Book on Indian Plants, Vols. I-III. Botanical Survey of India, Calcutta.

Nayar, M.P. 1996. Hot Spots of Endemic Plants of India, Nepal and Bhutan. Tropical Botanic Garden and Research Institute, Trivandrum.

Nayar, M.P. 1997. Biodiversity challenges in Kerala and Science of conservation Biology. In : P. Pushpangadan & K. S.S. Nair (eds.) Biodiversity of Tropical Forests the Kerala Scenario. STEC, Kerala, Trivandrum.

Prasad, N.S., 1998. Conservation Planning for the Western Ghats of Kerala. II. Assessment of habitat loss and degradation. Current Science 75: 228-235.

Ramesh, B.R., D. De Franceschi and J.P. Pascal 1997. Forest Map of South India – sheet: Thiruvananthapuram-Tirunelveli, Hors serie 22a. French Institute, Pondicherry.

Ramesh, B.R., P.V. Karunakaran and M. Balasubramanian 2003. Conservation Strategy for RET and Endemic Plant Species of Kerala. In: Proc. of the Workshop on Conservation and Research need of the Rare, Endangered and Threatened (RET) tree species in Kerala part of the Western Ghats. Kerala Forest Research Institute, Peechi. pp 13-17.

Sasidharan, N. 2003. Red Listed Threatened Tree Species in Kerala: A Review. In: Proc. of the Workshop on Conservation and Research need of the Rare, Endangered and Threatened (RET) tree species in Kerala part of the Western Ghats. Kerala Forest Research Institute, Peechi. pp1-9.

Sasidharan, N. 2004. Biodiversity Documentation for Kerala. Part 6: Flowering Plants. KFRI Handbook No. 17, Kerala Forest Research Institute, Peechi. 702p.

BIODIVERSITY IN "KAVUKAL" OF ONATTUKARA - AN OVERVIEW

S. Shilaja, T.N. Vilasini and Manoj Sebastian

Kerala Agricultural University,
Onattukara Regional Agricultural Research Station, Kayamkulam

Introduction

Onattukara is also known as Onadu and Onamoottukara. Onam means pleasure and prosperity. Pleasure is for the people and prosperity is obtained from agro ecological system existing in the area. Onattukara is surrounded by Thottappally spillway in the north, Neendakara in the south, Arabian sea in the West and Edanadu in the east. Onattukara comprises 42 panchayats and two municipalities with 7059 ha of land of which 6575 ha of land suitable for cultivation. Neendakara is a place where sea and lake join together. This is the reason why there is salt in the lake, fish population in this area is very high and there is ample scope for fish farming. Thottappally spillway in the north pushes the excess rain water along with the impurities to the sea and prevent the area from flood. Onattukara doesn't have direct access with sea. The area is suitable for cultivating a wide variety of crops like coconut, banana, root crops, tree crops, and fruit crops and lowland for growing rice, pulses, vegetables and oil seeds.

Majority of the cultivation in Onattukara are rainfed. The rain water received during rainy season was stored in Kulam (ponds) and fertile top soil was retained in kavukal maintained in each homesteads. Each homestead's were having three to four ponds to cater to the different needs of the family. In kavukal of Onattukara, there is a dense population of trees, shrubs, herbs, climbers etc. These kavukal were a true replica of forest, whatever service the forest played in conserving the environment, all these services were done here.

Structure and Biodiversity in Kavukal of Onattukara

If we examine the structure of kavukal, we can see trees growing at different heights, shrubs, herbs, climbers, orchids and rare species of plants along with a rich source of medicinal plants (Table 1). Among the trees, the biggest one will be of 30-50 m. height, below them are trees which grow up to a height of 20-30 m, then trees growing to height of 10- 20 m (Krishnan, 2000). Within these kavukal, there will be marshy land. Kavukal play a very important role in maintaining the ecosystem and conserving biodiversity. The micro climate, existing in kavukal help in increasing the availability of oxygen in the environment. The wide canopy and organic rich soil helps in conserving water and prevent soil erosion. The water conserved in these kavukal get collected in ponds and wells in the near by area. Earlier in Onattukara, for every one hectare of land, there were three to five ponds. The organic matter content obtained as a result of decaying of leaves, excreta of birds and animals living in kavukal improve the fertility status of the soil. Birds, animals and other living organisms protect the crop in the near by areas by eating the harmful insects and other organisms which in turn increase the productivity of crops. Kavukal also helps to raise the water level in the soil. The extensive root system of the trees in these kavukal prevents soil erosion, as a result of this water will remain in the soil which in turn raises the water level and trees also prevent wind.

Table 1. Plant species seen in Kavukal of Onattukara

	Scientific name	Family	Common name
1.	*Ficus glomerata* Roxb.	Moraceae	Athi
2.	*Ficus gibosa* Blume	Moraceae	Ethi
3.	*Ficus religiosa* L.	Moraceae	Arayal
4	*Hydnocarpus laurifolia* (Dennst.)	Flocourtiaceae	Marotti
5.	*Strychnos nux-vomica* L.	Loganiaceae	Kanjiram
6	*Alstonia scholaris* L. R. Br.	Apocynaceae	Ezhilampala
7.	*Holigarna arnotiana* Hook. f.	Anacardiaceae	Charu
8.	*Mimusops elengi* L.	Sapotaceae	Elangi

9.	*Calophyllum inophyllum* L	Cluceaceae	Punna
10	*Calophyllum vettiyam* L.	Cluceaceae	Cherupunna
11.	*Aegle marmelos* L	Rutaceae	Koovalam
12	*Saraca indica* L	Caesalpineaceae	Ashokam
13.	*Spondias mangifera* Willd.	Anacardiaceae	Ambazham
14.	*Nerium indicum* Mill.	Apocyanaceae	Arali.
15	*Anaphyllum wightii* Schott	Araceae	Keerikizhangu
16.	*Sida* spp	Malvaceae	Kurumthotti
17.	*Scoparia dulcis* L.	Scrophulariaceae	Neenakani
18.	*Myristica malabarica* Warb.	Myristicaceae	Kattujathi
19.	*Bombax ceiba* L.	Bombacaceae	Elavu
20	*Calamus rotang* L.	Palmae	Chooral
21	*Abrus precatorius* L.	Papilionaceae	Kunni
22.	*Adenanthera pavonina* L.	Mimosaceae	Manchadi
23	*Erythrina indica* Lam. Syn. *E. varigata* var. *orientalis*	Papilionaceae	Murikku
24.	*Terminalia paniculata* Roth.	Combertaceae	Maruth
25.	*Dalbergia latifolia* Roxb.	Papilionaceae	Eetti
26	*Sterculia urens* Roxb.	Sterculiaceae	Thondi
27.	*Carallia brachiata* syn. *C. lucida*	Rhizhophoraceae	Vallabham
28	*Cinnamomum sulphuratum* Nees	Lauraceae	Vazhana
29.	*Borassus flabelifer* L.	Palmaceae	Karimpana
30.	*Acacia intsia* Wight & Arn.	Mimosaceae	Inja
31.	*Pinus wallichiana*	Pinaceae	Pine
32.	*Azadirachta indica* A. Juss.	Miliaceae	Aryaveppu
33.	*Buchanania lactifolia* Wight	Anacardiaceae	Kolamavu
34.	*Ficus bengalensis* L.	Moraceae	Peral
35.	*Garcinia gummigutta* (L.) Rob	Clusiaceae	Kudampuli
36.	*Odinia woodier* Roxb.	Anacardiaceae	Udi
37.	*Mallotus philippensis* (Lam.) Mueli.	Euphorbiaceae	Kurangumanjal
38.	*Semicarpus auriculatus* Beed	Anacardiaceae	Vellacharu
39.	*Syszygium cargophyllatum* (L.) Alstan	Myrtaceae	Njara
40.	*Santalum album* L.	Santalaceae	Chandanum
41	*Vacteria indica* L.	Dipterocarpaceae	Vellapayin
42.	*Achyranthus aspera* L.	Amaranthaceae	Kadaladi

43	*Aerva lanata* (L.) Juss.	Amaranthaceae	Cherula
44.	*Andrographis paniculata* (Burm. f)	Acanthaceae	Kiriyath
45.	*Callicarpa tomentosa* (L.) Murvay	Verbenaceae	Cheruthekku
46.	*Clerodendron viscosum* Vent.	Verbenaceae	Peruvalam
47.	*Ixora coccinea* L.	Rubiaceae	Kattuthechi
48.	*Jasminum* sp	Oleaceae	Kattupichi
49.	*Mussanda bellila* Buch.Ham	Rubiaceae	Vellilathali
50.	*Myxopyrem serratulum* Hill	Oleaceae	Chathuramulla
51.	*Narangamia alata* Wight	Meliaceae	Nilanarakam
52	*Biophytum sensitivum* (L.)	Oxalidaceae	Mukkutti
53	*Cardiospermum halicacabum* L	Sapindaceae	Uzhinja
54.	*Centella asiatica* (L.) Urban	Apiaceae	Kudangal
55	*Curculigo orchioides* Gaertn.	Liliaceae	Nilapana
56.	*Gymnema sylvestre* (Retz.)	Asclepidiaceae	Chakkarakolli
57.	*Ocimum sanctum* L.	Labiatae	Thulasi
58.	*Piper longum* L	Piperaceae	Thippali
59.	*Rauvolfia serpentina* Benth	Apocynaceae	Sarpaganthi
60.	*Anarmirta cocculus* (L.) Wight	Menispermaceec	Nanjuvalli
61.	*Asparagus recemosus*	Liliaceae	Sadaveri
62.	*Calicopteris floribunda* Lam.	Combretaceae	Pullanni
63.	*Dioscorea* sp	Dioscoreaceae	Kattukachil
64.	*Gloriosa superba* L.	Lilliaceae	Menthonni
65.	*Piper nigram* L.	Piperaceae	Kurumulaku
66.	*Tylophora asthamatica* Wight & Arn.	Asclepidaceae	Vallipala
67.	*Tinospera cordifolia* Miers	Menispermaceae	Amrutu
68.	*Clitoria ternatia* L.	Fabaceae	Sangupushpam

Aquatic plants seen in kavukal of Onattukara

1.	*Nelumbium speciosm*	Nymphaceae	Lotus
2.	*Nymphea esculenta*	Nymphaceae	Water lily
3.	*Bacopa monnieri*	Scrophuleriaceaea	Brahmi
4.	*Salvinia molesta*	Salveniaceae	Payal
5.	*Eichornia* sp.	Pontederiaceae	Kulavazha

Earlier, people were so attached to nature and trees were considered as sacred. They also worshipped snakes for their own existence.

They were afraid of snakes, so as part of existence they kept apart a piece of land for kavu and started worshipping snakes. Kavukal originated in homesteads in order to worship snakes, and these kavukal were the dwelling place of poisonous snakes and prevent them from roaming about in the homesteads. Kavukal were considered as a sign of prosperity. People had taken special interest in the maintenance of kavukal. They also considered kavukal as a true representative of nature. All the poisonous and cruel animals depend on these kavukal for their existence, so these kavukal were considered as "pranente prakrithi". Culture and customs of our fore fathers were having an intimate relationship between the three factors *viz.*, snakes, human beings and environment. Kavukal is a typical example of this relationship. Kavukal are seen attached to temples also, the only difference is in the size.

Believes associated with kavukal

Various believes existing in the community in connection with Kavukal. There is a saying "Kavukal Theendaruthu Kulangal Vattum", and "Kavil Irumpu Kondu Kayararuthu". This is to prevent destruction of trees in the kavu, which plays a very important role in the conservation of soil and water. A single Anjali tree itself is a habitat, many organisms depends on this tree for their existence. Birds depend on fruits, insects and some organisms feed on leaves and birds build their nest on trees.

Medicinal plants in kavukal of Onattukara

Kavukal is a store house of medicinal plants. Hundreds of species of medicinal plants are seen in these kavukal. Our ayurvedic culture has got a profound influence in the origin of kavukal. Earlier, landlords and Namboodiris kept apart large piece of land for growing trees and other plant species of medicinal value. This is a clear evidence of their knowledge on medicinal properties of plant species. Most of their medicinal needs were met from the plants maintained in these kavukal (Table 1). When ayurveda has developed in a big way, people started using plants in kavukal and some of them are in the stage of extinction. Plants found in the natural habitat were considered to have high medicinal value

Birds and animals in kavukal of Onattukara

Kavukal is a dwelling place of birds, animals and reptiles. A wide variety of birds live in kavukal (Table 2). These birds play a very important role in the conservation of environment. Birds and plants in these kavukal are mutually benefited for their survival. Majority of them depends on fish in the ponds attached to the kavu and they also feed on insects and similar organisms in the paddy fields. They help in the dispersal of seeds, pollination, addition of organic matter through their excreta and protecting the crop from harmful insects (Suresh, 2000).

Table 2. Birds seen in kavukal of Onattukara

Scientific name	Common name	Malayalam name
Psycnonotus cafer	Redvented bulbul	Nattubulbul
Dicrurus luerulescens	Racket tailed drongo	Kattumuzhikki
Pycnonotus jacosus	Red whiskered bulbul	Erattaunjali
Acridotheres iristis	Common myna	Nattumyna
Acridotheres fuscus	Jungle myna	Kinnarimaina
Turloides striatus	Jungle babbler	Kareelakili
Egretta intermedia	Median egret	Edakokku
Dinopium benghalense	Goldenbaked wood pecker	Maramkothi
Pycnonotus jacosus	Red whiskered bulbul	Erattajnali
Corvus splendens	Hose crow	Chenakakka
Athene brama	Sppotted owlet	Pullinathu
Dendrocitta vagabunda	Indian treepie	Olenjali
Otus bakkamoen	Collared scops owl	Cheviyan nattu
Ardeola grayil	Pond heron	Kulakokku
Halcton smyrnensis	White dreasted king fisher	Ponman
Anhinga rufa	Snake bird	Neerkakka
Centropus sinensis	Crow-pheasant	Uppan
Copsychus saularis	Magpie robin	Mannathipullu
Columba livia	Blue rock pigeon	Ambalapravu
Ploceus phillippinus	Baya sparrow	Thukkanumkuruvi
Haliastur indus	Brahminy kite	Krishnaparunthu
Amaurornis phoenicurus	White breasted waterhen	Kulakozhi

A wide variety of snakes, lizards, mangoose, monkey, squirrels, bats amphibians, reptiles etc lives in Kavukal. Kavukal serve as a dwelling place for these animals and protect them from enemies. These animals depends on small animals, birds, insects in the Kavukal for food and they in turn helps in the existence of Kavukal through pollination, dispersal of seeds and addition of organic matter through their excreta which in turn improve the fertility status of the soil.

Conclusion

Our fore fathers were aware that environmental equilibrium was maintained by plants, animals rivers sea, mountains, wind and change in any one of the factors will have a detrimental effect in all other factors. This was the main reason why they loved nature and they considered these factors as sacred and utilized them with out creating any damage (Pradeep, 2000). Kavukal is a typical example of this. Many believes existed in this community in association with kavukal to protect them from destruction. Now this situation has changed, with the increase in population, and change in the attitude of the people they converted kavukal in to cultivable land. The trees in these kavukal were cut and removed for constructing houses, temples etc. Now a days in order to convert the kavukal for various other purposes the deities in these kavukal were taken to near by temples. In this era of democratic decentralization people must be made aware of the importance of maintaining kavukal and their role in conserving biodiversity conservation of soil, water, storehouse of medicinal plants and dwelling place of birds and animals etc. In Onattukara where the soil is sandy with low water holding capacity low fertility status, destruction of kavukal and Kulangal, which helped in soil and water conservation, improving the fertility status of the soil through addition of organic matter resulted in flood during rainy season and drought in summer season.

References

Krishnan, P.N. 2000. 'Keralathilae Sarpakkavukal' *Swadesi science*. 10(1): 5-12

Pradeep, K.R.J. 2000. Paristhithinasavum samskarachuthiyum thammilulla bandathepatty oru anveshanam. *Keralathinte marunna mukhachaya*. Kurushekthra Prakasan Pvt. Ltd. Convent road, Cochi. 317-322

Suresh, M. 2000. Kavukalum anthareeksha paripalanavum. *Sarparadhanayum prakruthy samrakshanavum*. Current Books. Kottayam: 123-134.

EFFECT OF COIR RETTING ON MARINE FUNGAL FLORA

K. Raveendran

P.G. Dept. of Botany, Sir Syed College, Taliparamba, Kannur, Kerala

Abstract

To study the effect of coir retting on the distribution of marine mycoflora, three coir retting fields adjoining to three estuaries of Northern coastal Kerala were selected and studied. Intertidal wood substrates were collected from both coir retting fields and from the three adjoining estuaries. Marine fungi were isolated and identified.

Altogether 58 species of marine fungi were obtained. Of these 40 ascomycetes, 1 basidiomycete, 14 mitosporic fungi and three were unidentified ascomycetes recorded. Seven species of marine fungi were present in all the six collection sites. 10 marine fungi were present in all the three estuaries and 10 were found in two estuaries while 30 were present in any one site, in such a way that there was no redundancy of any one species.

Maximum marine fungal species diversity was found in Tellicherry estuary (50 species) while all the three coir retting fields showed remarkably lower fungal diversity (7 species). This lower fungal diversity in the coir retting fields might be due to the effect of phenolic compounds, the lesser amount of dissolved oxygen and the particulate matter accumulated during the process of retting.

The common marine fungi encountered in al the six water bodies were *Aniptodera Chesapeakensis, Halosarpheia marina, Marinosphaera mangrovei, Lulworthia grandispora, Lulworthia* sp., *Halocyphina villosa* and *Cirrenalia pygmea.*

Introduction

Marine fungi have been intensively studied over the past 50 years. These are very interesting and fascinating groups of organisms. The spores of many of the marine fungi are very distinctive. The role of marine fungi in the break down of organic matter in the sea and adjoining coastal waters are well documented (Kohlmeyer & Kohlmeyer, 1979). In addition it is possible that some of them produce antibiotics and other useful metabolites. This young branch of biology, the marine mycology, has a rapid growth during these decades. More than 600 marine fungi have been described to this date and the number is increasing almost day by day. A perusal of the literature revealed that the south west coast of India remained virtually unexplored. Very little is known about the marine fungi of Kerala coast. A few earlier workers (Becker & Kohlmeyer, 1958, Kohlmeyer, Schmidt & Nair, 1967, Prabhakaran, Gupta and Krishnankutty, 1987, Prasannarai & Sridhar, 2001, 2003, Maria & Sridhar, 2002 and Raveendran, 2003) have made some occasional collections of marine fungi from some locations in the Kerala coast. This paper present the fungal diversity of three estuaries and three adjoining coir retting fields and the effect of coir retting on marine fungi.

Materials and Methods

Collection and treatment of wood samples

Twenty four collections were made from Tellicherry estuary, 4 collections were made from Mahe A & Kadalundi A, while five collections were made from Dharmadam B, four from Mahe B and three from Kadalundi B. Collections of wood samples were made during monsoon, post monsoon and premonsoon. In each season, at least 3 collections were made from each location. Collections were made from 1992 to 1994. Wood samples were collected from the intertidal zone of the collecting location and total of 659 intertidal wood were collected during the study (Tellicerry 327, Mahe A 135, Kadalundi A 78, Dharamdam B 48, Mahe B 37 and 34 from Kadalundi B). The wood samples collected included decaying wood substrates of mangroves and other unidentified wood substrates. Among the 659 wood samples collected, 247 were collected during monsoon, 208 were collected during post monsoon and the rest 204, during premonsoon. Uniformity in the size of the wood samples were maintained to a certain extent by picking only samples that came with in the size range of 7.5 – 10 x 1 – 2 cm. Collected wood samples were washed well and placed in sterile polythene bags and brought to the laboratory on the same day. The wood samples were subjected to preliminary screening with in a week after collection, and marine fungi were isolated. The wood samples were incubated in sealed sterile polythene bags in the laboratory at room temperature. The incubated wood samples were periodically examined for marine fungi and the fungi observed were isolated and recorded. These periodical examinations were continued up to six months. The number of species obtained from each wood sample was also recorded.

Isolation of marine Fungi

The wood samples were examined for sporulating structures such as conidia, perithecia, pycnidia and basidiomata under a stereomicroscope. Fruiting structures were removed with a fine pair of forceps or a needle. The isolated fungi were examined under compound microscope, Free hand section, or in case of small, delicate specimens, cryostat sections were made. Spores and sections were always mounted in the first instance in seawater so that any appendages present can dilate and their true morphology can be studied. Permanent mounts were made with lacto phenol-cotton blue and sealing the cover glass with DPX.

Drawings were prepared using a mirror type camera lucida.

Table 1. Occurrence of species of marine fungi at different locations in estuaries and coir retting fields

Name of the fungi	Estuaries			Coir retting fields			No. of sites the fungus occurred
	Tellicherry A	Mahe A	Kadalundi A	Dharmadam B	Mahe B	Kadalundi B	
Ascomycetes							
Aigialus grandis Kohlm & Schatz	√						1
Aigialus mangrovie Borse	√						1
Aigialus parvus Kohlm & Schatz	√						1
Aniptodera chesapeakensis Shearer & Miller	√	√	√	√	√	√	6
Aniptodera longispora Hyde	√						1
Aniptodera mangrovei Hyde in Hyde Farrant & Jones	√						1

Aniptodera haispora Vrijmoed, Hyde & Jones	√						1
Aniptodera salsuginosa Nakagiri & Ito	√						1
Aniptodera sp I	√						1
Aniptodera sp II		√					1
Biatriospora marina Hyde & Borse	√						1
Ceriosporopsis halima Linder in Barghoom & Linder	√						1
Dactylospora haliotrepha (Kohlm & Kohlm) Hafellner	√	√	√				3
Dryosphaeria navigans Koch & Jones	√						1
Halosarpheia abonnis Kohlm.	√	√	√				3
Halosphaeria hamata (Hohnk) Kohlm.	√	√	√				3
Halosarpheia marina (Cribb & Cribb)Kohlm	√	√	√	√	√	√	6
Halosarpheia minuta Leong, Tan, Hyde & Jones	√	√					2
Halosarpheia ratnagiriensis Patil & Borse	√	√					2
Halosarpheia retorquens Shearer & Crane	√						1
Halosarpheia viscosa (I. Schmidt) - Shearer & Crane ex Kohlm & Volkm – Kohlm	√						1
Halorosellinia oceanica (Schatz) Whally, E.B.G.Jones, K.D.Hyde & T.Leassoe		√	√				2
Leptosphaeria australiensis - (Cribb & Cribb) G.C. Hughes	√	√	√				3
Lignincola laevis Hohnk	√	√	√				3
Lignincola longirostris (Cribb & Cribb) Kohlm	√						1
Lignincola tropica Kohlm	√						1
Lindra hawaiiensis Kohlm & Volkm Kohlm	√	√					2

Lulworthia grandispora Meyers	√	√	√	√	√	√	6
Lulworthia sp I	√	√	√	√	√	√	6
Lulworthia sp II	√	√	√				1
Manglicola sp			√				1
Marinosphaera mangrovei Hyde	√	√	√	√	√	√	6
Ocostaspora apilongissima Jones, Johnson & Moss	√						1
Payosphaeria minuta Leong in Leong, Tan, Hyde & Jones	√						1
Pleospora pelagica Johnson	√		√				2
Savoryella lignicola Jones & Eaton	√	√	√				3
Savoryella paucispora (Cribb & Cribb) Koch	√	√	√				3
Savoryella sp I			√				1
Salsuginea ramicola Hyde	√						1
Unidentified ascomycetes II							
Unidentified ascomycetes III	√		√				2
Unidentified ascomycetes V	√		√				1
Verruculina enalia (Kohlm.) Kohlm et Volkm - Kohlm.	√	√	√				3
Basidiomycete	√	√	√				
Halocyphina villosa Kohlm & Kohlm	√	√	√	√	√	√	6
Mitosporic fungi	√	√	√	√	√	√	
Cirrenalia basiminuta Raghukumar et Zainal in Raghukumar, zainal et Jones	√						1
Cirrenalia fusca Schmidt			√				1
Cirrenalia macrocephala (Kohlm) Meyers et Moore	√		√				2
Cirrenalia pygmea Kohlmayer	√	√	√	√	√	√	6
Cladosporium sp. (83)			√				1
Dendryphiella salina (Sutherland) Pugh & Nicot	√						1

Periconia prolifica Anastasiou	√	√	√				3
Phialophorophoma litoralis Linder in Barghoorn & Linder	√	√					2
Phoma sp.	√						1
Trichocladium alopallonellum (Meyers & Moore) Kohlm & Volk Kohlm	√	√	√				3
Trichocladium sp.		√					1
Zalarion maritimum (Linder) Anastasiou	√	√					2
Zalarion varium Anastasiou	√		√				2
Zalerion sp	√						1
Total number of species obtained	50	26	27	7	7	7	

Results and Discussion

Floristic account

A total of 58 marine fungi were obtained during the course of this study. Out of the 58, forty were Ascomycetes, one was a basidiomycete and 14 were mitosporic fungi and 3 were recorded as unidentified Ascomycetes. Among these 58; 18 species recorded by Prasannarai and Sridhar (2001) from Bekkal and Kovalam while all other were reported for the first time from the Kerala coastal waters by Raveendran (2003). 7, out of the total 58 species of marine fungi were present in all the six collection sites, 10 were present in all the three estuaries, 10 were found in two estuaries while 28 were present in anyone site in such a way that there was no redundancy of any one species. Among the 58 marine fungi, only seven were present in both ie. Estuaries and in coir retting fields. Among the six collection sites, maximum marine fungal species diversity was found in Tellicherry estuary (50 species) while all the coir retting fields showed remarkably lower fungal diversity (7 species). This lower fungal diversity in the coir retting fields might be due to the effect of phenolic compounds, the lesser amount of dissolved oxygen and the particulate matter accumulated during the process of retting. Interestingly, during monsoon, in all the retting fields the species diversity were more and less during premonsoon. The same was reported by Remani *et al.* (1988) during their study about the effect coir retting on the estuaries flora and fauna. During the study period, 39 fungal species were obtained during monsoon, 42 during post monsoon and 38 during pre monsoon and hence it seemed the seasonal variation does not have much influence upon the marine fungal flora of estuaries. While in the coir retting fields the maximum number of species occurred during monsoon season and less during premonsoon. It is observed that the majority of the marine fungi collected during the present study belonged to Ascomycota. The reports of several previous workers (Kohlmeyer, 1984, Kohlmeyer and Volkmann Kohlmeyer, 1987, Hyde, 1988a, Sarma & Vittal, 2001, Raveendran, 2003) support the same observation. The high diversity observed in the present study is similar to the reports about other tropical locations (Borsem, 1988, Hydem, 1988a, Hyde & Jonesmm, 1989a, Sarma, Hyde & Vittalm, 2001, Prasannarai & Sridharm, 2001 and Raveendran, 2003).

Acknowledgement

The author is greatly indebted to Dr. P. Manimohan, Reader, Dept. of Botany, Calicut University and to Dr. K.M. Leelavathy, Former Professor, Dept. of Botany, Calicut University for their continuous encouragement, guidance and support.

References

Becker, G. and Kohlmeyer, J. 1958. Deterioration of wood by marine fungi and its special significance for fishing crafts. *Journal of the Timber Dryers and Preserver's Association of India.* 4: 1-10.

Borse, B.D. 1988. Frequency of occurrence of marine fungi from Maharashtra coast, India. *Indian J. Mar. Sci.* 17: 165-167.

Hyde, K.D. 1988. Studies on the tropical marine fungi of Brunei. *Bot. J. Linn. Soc.* 98: 135-151.

Hyde, K.D. and Jones, E.B.G. 1989a. Ecological observation on marine fungi from the Seychelles. *Bot. J. Linn. Soc.* 100: 237-254.

Kohlmeyer, J. 1984. Tropical Marine fungi. *Mar. Ecol. P.S.Z.N.I.* 5(4): 329-378.

Kohlmeyer, J. and Kohlmeyer, E 1979 *Marine mycology - The higher fungi* Academic press, New York, USA.

Kohlmeyer, J., Schmidt, I. and Nair, N.B. 1967. Eine neue *Corollospora* (Ascomycetes) aus dem Indischen Ozean und der Ostsee. *Ber. Dtsch. Bot. Ges.* 80: 98-102.

Kohlmeyer, J. and Volkmann-Kohlmeyer, B. 1987a. Marine fungi from Belize with a description of two new genera of ascomycetes. *Bot. Mar.* 30: 195-204.

Kohlmeyer, J. and Volkmann-Kohlmeyer, B. 1987b. Marine fungi from Aldabras, the Galapagos and other tropical islands. *Can. J. Bot.* 65: 557-582.

Maria, G.L. and Sridhar K.R. 2002. Richness and diversity of filamentous fungi on wood letter of mangroves along the west coast of India. *Curr. Sci. Vol.* 83. No. 12: 1573-1580. Prabhakaran, N., Gupta, R. and Krishnankutty, M. 1987. Fungal activity in Mangalvan: an estuarine mangrove ecosystem. In: *Proceedings of the National seminar on Estuarine Management* (ed. Nair. N.B.) STEC, Kerala, pp. 458-463.

Prasannarai, K. and Sridhar, K.R. 2001. Diversity and abundance of higher marine fungi on woody substrates along the west coast of India. *Curr. Sci.* 81(3): 304-311.

Prasannarai, K. and Sridhar, K.R. 2003. Fungal assemblage and diversity on periodically sampled intetidal woody litter. *Indian J. Mar. Sci. Vol.* 32(4): 329-333.

Raveendran, K. 2003. Diversity of marine fungi of Kerala coastal waters. National symposium on prospecting of fungal diversity and emerging Technologies. Abstracts PP 24.

Sarma, V.V., Hyde, K.D. and Vittal, B.P.R. 2001. Frequency of occurrence of mangrove fungi from the east coast of India. *Hydrobiologia* 455: 41-53.

Sarma, V.V. and Vittal, B.P.R. 2001. Biodiversity of manglicolous fungi on selected plants in the Godavari and Krishna deltas, east coast of India. *Fungal diversity* 6: 115-130.

A LITTLE KNOWN USE OF THE PALMYRA PALM: A SHORT RANGE GRAVITY-FLOW IRRIGATION SYSTEM FROM TRADITIONAL AGRICULTURE

V.S. Ramachandran

Division of Forest Ecology & Biodiversity Conservation,
Kerala Forest Research Institute, Peechi, Kerala - 680 653, India.

Abstract

Next to coconut palm, the palmyra palm (*Borassus flabellifer* L.) is of great economic importance in the village life in South India. It provides wood for construction, leaves for thatching, is a source of jaggery and is also put to use in a variety of situations. The paper describes a traditional irrigation system from Palghat District of Kerala, exclusively utilizing the palmyra palm that has not been documented so far. Irrigation utilizing the gravity flow technology and lift irrigation used to be the conventional systems in Kerala, for irrigating the paddy fields. Rivers, streams, ponds or lakes are involved in these systems and where ponds and lakes are involved, water used to be taken to the crop fields by cutting the bund of the pond/lake on one side. The irrigation system using barrels made of the palmyra palm is a highly improvised gravity flow system that enabled the farmer to tap the water as and when required, much the same way as in a pipe and tap system. The most important component of the system is a branched conduit work made of palmyra palm or the '*karim-pana*' (Malayalam). The paper describes the salient features of this irrigation system.

Introduction

Palms are one of the versatile monocots of the tropical and subtropical regions of the world, of which about 200 genera are currently recognized. India has to its credit about 100 species of palms falling under 21 genera; they are distributed in the three major geographical regions, *viz.*, Peninsular India, Northeastern India and Andaman and Nicobar Islands (Renuka, 2001).

The genus *Borassus* is one of the widely distributed palms, extending its range in a broadband from western Africa to eastern Indonesia (Davis and Johnson, 1987). Species of the genus are commonly cultivated in India, Southeast Asia, and occasionally in other warm regions of the world, including Hawaii and southern Florida (Morton, 1988).

Borassus flabellifer L. the palmyra palm, which thrive well in the tropical and subtropical regions (Kamble, 2003), is a highly resourceful tree to mankind of which no part is wasted. The species exists both in wild as well as in cultivation, ranging from sea level up to 760 m (Fischer, 1931). In India, the cultivated populations are widespread in Kerala, Tamilnadu, Karnataka, Andhra Pradesh, Maharashtra, Orissa, Madhya Pradesh, Bihar and West Bengal. Isolated patches are also seen in Assam, Gujarat, and Uttar Pradesh (Anonymous, 1948).

The palmyra palm is a monocaulous solitary dioecious palm reaching a height of 25-30 m and an average diameter of 50-60 cm. The young trunk covered with clasping leaf bases transforms to smooth bole with narrow leaf scars at age and is black in colour. The large fan-like lobed leaves clustered at the tip of the trunk, make a loose crown and the leaf bases clasp the stem firmly. Inflorescences are interfoliar; male inflorescence is with stout cylindrical branches while female inflorescence is sparingly branched. Fruits are semiglobose to globose and deep brown to black when ripe.

Palmyra palm has been an inseparable part of human culture and tradition since ancient times; each and every part of the tree is useful in a way or the other. The palm yields timber, thatching material, fibre of various kind, drinks, sugar, famine food, firewood, and many folk medicines. All parts of the tree like immature wood, leaves, petiole, inflorescence, and soft useless part of the stem are a good source of fuel, when dry; the immature wood is specially used as fuel in brick kilns.

The mature trunk of the palmyra palm is very durable and valuable. The hard outer wood is used for constructing frame work of roofs, house-hold posts and pillars, furniture and other utensils like eaves troughs and spouts (Morton, 1988), ruler, walking sticks, etc; the wood is also of popular use for furniture, grilles for windows and stair cases.

In addition to the conventional uses of the palm trunk mentioned above, Fischer (1931) mentioned that the stem is used as water pipes and troughs, though no documented report of its utility in irrigation or other water transport systems is available till date. Very recently I have come across a hitherto undocumented use of the palm in a traditional water engineering system, in a village namely Vilayannur from the Palakkad District of Kerala, and I document it here.

A glimpse through the history of irrigation

In the evolutionary history of the human society, cultivation of food crops in large quantities necessitated irrigation. Prehistoric Egyptians (*c.* 5,000 BC) diverted water from river Nile through channels for use during periods of water scarcity. It appears that the technology spread to other regions of the world and reached the Indian sub-continent by *c.* 3,000 BC (Encyclopedia Brittanica, 1974). This enabled large-scale irrigation possible. The narrow mouth of the channel used to be plugged by stones, soil and twigs, whenever irrigation was not required. When desired, the plugs are removed and water permitted to flow through and diverted to fields located at different points in the course of the channel. Such irrigation systems used the gravity flow of water along contours.

Check dams provided a major improvement in the harvest of water, in the sense that water flowing through the river is collected in the reservoir and diverted to crop fields through a dug out distributory system comprising large and narrow channels (Bagchi, 1995). The reservoir being a storage device, water harvest and delayed use inherent in the system are taken advantage of.

Before the emergence of large check dams, people in India used to harvest water using small lakes and ponds and lift-irrigation used to be the prevalent means of short-range irrigation. Water was drawn from well, pond or any other source by various devices operated manually or using animals. The lifted water was transported through conduits to the crop fields. The conduits were dug in soil or made out of palm stems. Either naturally occurring ponds or dug out artificial lakes were used for collecting water, during the monsoon. When the monsoon recedes and scarcity of water begins, one of the bunds of the lake is broken and water channelised to the fields.

The undocumented irrigation system using the palmyra palm

Like other villages of Palakkad District, the Vilayannur Village is basically agrarian with vast extent of rice fields. Different varieties of rice differing in lifecycle from 90 to110 days are cultivated here. Back in history, farmers depended on monsoon showers and small water bodies like ponds or lakes for irrigating the crop. When the fields happened to be at a lower level than the pond, water was tapped by breaking the bund a little on one side; this procedure always had inherent difficulties, as the outlet frequently went out of control. Also, the broken bund had to be closed every time, for tapping the water in instalments, and needed to be closed again, in order to get the pond filled during the rainy season. However, some of the thoughtful farmers seem to have used a novel technique in tapping the pond water in instalments, as and when needed, using a pipe system made of palmyra palm, without involving the breaking of the bund. The details of the system are as follows.

The irrigation system basically comprises of a pond (*c.* 0.5 ha in area) that collects water during monsoon showers. It is shallow on one side (1.5 m), deeper on other sides (2.5-3 m), deepest at the centre (*ca.* 4 m) and a 3-4 m deep well occupies the centre of the pond. Two settlements and a farmhouse campus constitute the watershed of the pond. Monsoon fills the pond and the water collected during the receding Southwest monsoon (October-November) is used for irrigating the second crop of rice, the coconut in homesteads, banana plantation and vegetable gardens between January and June, till the arrival of the next Southwest monsoon (*kala varsham*). The topography of the area is such that the farmhouse gradually slopes into the pond and subsequently to the rice fields. This relief feature is cardinal of the working system and the bund of the pond at the sloping edge is kept two to three meters thick, so that it withstands the pressure of the water.

The pipe system comprises of several *pipelets* (hollow cylinders of 50-60 cm long) made from the trunk of palmyra palm, linked together, and it consists of three units : (1) A *horizontal channel* - 20-30 m in length and laid underground, one end of which is placed underneath the basin of the pond and the other end outside the pond. (2) An *inner vertical branch* - connected to the blind end of the horizontal channel below the basin of the pond, with its mouth into the pond. (3) An *outer vertical branch* - connected to the horizontal channel at its extremity and opening into the rice fields. Some times the horizontal channel is also supplemented with an *additional outer vertical branch* attached somewhere in the middle.

The inner vertical branch opening into the pond is elevated than the outer vertical branch(s) and consists of a few pipelets placed one above the other, of which four are removable. The inner vertical branch is situated in a depression in the pond, three sides of which is fortified with kiln bricks so as to prevent sliding of the mud wall, while the fourth side remains open to maintain the flow of water. Water drains from the pond to the inner vertical branch, passes through the horizontal channel and comes out through the outlet(s), the outer vertical branch(s), situated in the rice field. The outer vertical branch (s) is made of 2 or 3 pipelets attached one over the other with cementing material. Both the inner and outer vertical branches are closed with a lid, usually made of granite, placed over them. The lid of the inner vertical branch acts as a valve where as that of the outer vertical branch (s) prevents siltation into the pipe system while not in use. The horizontal channel is formed of many pipelets joined in a straight-line array using some crude cementing material. The cementing material is made of lime, sand and gravel. In the Malayalam vernacular, the pipelets are known *'panayural'* (*pana* = palmyra palm, *ural* = mortar) and the lid, *'moodikallu'* (*moodi* = lid, *kallu* = stone).

How the system works?

When the system is not in use, the vertical branches (both inner and outer) are closed with the lids and the gap with the lid(s) is sealed with the help of hay or grass. The pipelets of the inner vertical branch are movable so that water can be drained from the pond in a controlled manner, stage by stage, as with a tap. When the lid over the inner vertical branch is removed, the amount of water available in the pond above the mouth of this pipelet can be drained for irrigation. When the water level of the pond reaches the level of the uppermost pipelet in the inner vertical branch, the intake of water through the system stops. If more water is required, the next pipelet is removed so that water for a level further is released from the pond. In this way the pond can be emptied completely to its bottom, stage by stage. Such a system suffices to cater to the requirements of a rice field about 5-6 ha for two crops a year.

Advantages of the system

In olden days when other farmers were cultivating only single crop of rice a year, this farmhouse with the traditional irrigation system described in this paper, could harvest two or three crops year using water in the pond. This was possible because water could be extracted from the ponds in installments, as and when required. The system worked more or less like a tap system, which permitted control of water extraction. However, the system was not widely used by farmers, perhaps the technical know how was limited to a few individuals. Even today, the irrigation system using palmyra palm could be made use of, where the agricultural farms/rice fields form a slanted landscape, and where there is scope for digging up a pond. Although the initial cost of organizing the irrigation system would be a little high, this would be a long-term investment to the cause of a sustainable water harvest and successful agriculture.

The pipe system made of pipelets is the most important part of the system; being made of palmyra palm it is highly resistant to degradation in wet soil and water. The design is quite remarkable, when we realize that the system was developed, when modern engineering materials like cast iron, cement concrete and PVC were not available. Today, for translocation of liquids like water, oil and gas, pipes made of clay, cast iron, aluminium, PVC, cement concrete, etc. are used. The three-generation-old (*c.* 100 year) pipe system is still found to be healthy.

With the changing lifestyles, more and more of indigenous and traditional knowledge are getting lost forever. Synthetic and health hazardous materials replace conventional commodities of day-to-day life. Judicious use of traditional and ethnic knowledge might help evade many unwanted synthetic materials from our daily life. A conscious effort to document, assemble and explore the potentials of traditional/ethnic knowledge alone can resolve the situation.

Acknowledgement

The author is thankful to Dr. J.K. Sharma, Director, Kerala Forest Research Institute, as a continuous source of inspiration. Comments and suggestions offered by Dr. K. Swarupanandan, on an earlier draft of the MS are highly appreciated.

References

Anonymous, 1948. *The Wealth of India: raw materials*. Vol. 1. Publication Division, CSIR, New Delhi. 203-207 pp.

Bagchi, K.S. 1995. *Irrigation in India: History and Potential of social management*. Upalabdhi, Trust for developmental initiatives, New Delhi.

Davis, T.A and D.V. Johnson 1987. Current utilization and further development of the palmyra palm (*Borassus flabellifer* L. Arecaceae) in Tamil Nadu state, India. *Economic Botany* 41 (2): 247-266.

Encyclopedia Britannica Inc. 1974. Agriculture. Pp. 324-366. *In: The new encyclopedia Britannica, Macropedia,* Vol. 1. Helen Hemingway Benton, Chicago.

Fischer, C.E.C. 1931. Family Palmaceae. Pp. 1561-1561. *In*: Gamble, J.S. *Flora of the Presidency of Madras*, Vol. 3: Adlard & Sons, London.

Kamble, K.D. 2003. Palm gur industry in India. *Indian J. Traditional Knowl.* Vol. 2(2): 137-147.

Morton, J.F. 1988. Notes on distribution, propagation and products of *Borassus palms* (Arecaceae). *Economic Botany*, 42(3): 420-441.

Renuka, C. 2001. Palms of India: Status, threats and conservation strategies. Pp. 197-209. *In*: Uma Shaanker, R., K.N. Ganeshaiah and K.S. Bawa (Eds.). *Genetic Resources: Status, threats and conservation strategies.* Oxford & IBH, New Delhi.

[illegible]

[illegible]

[illegible]

[illegible]

[illegible]

[illegible]

[illegible]

MEDIA DEVELOPMENT FOR ANAEROBIC PROTOZOAN-BACTERIAL CONSORTIUM AND EFFECTIVE USE

V.K. Madhu

Research faculty, Amrita Vishwa Vidyapeetham
[Deemed University], Coimbatore.

Abstract

The study of microbial diversity leading to the preservation of mixed communities, consortia, is important to solve new and emerging problems and to advance biotechnology. The area addresses the need of preserving organisms, including *in-situ* preservation as well as functionally efficient consortia so that human society can use it effectively. It has been reported that syntrophic relation between anaerobic protozoans and anaerobic bacteria were found accelerating degradation rates when compared with prokaryotic consortium. But their significance in wastewater treatment plants is less fully under stood. Hence culturing of anaerobic protozoans became significant, as it is the preliminary step for studying their role in biodegradation and biochemical pathways as consortia with prokaryotes in wastewater treatment. Hence the objective of the work was to develop suitable culture techniques for the growth of protozoans isolated from anaerobic reactor sludge [buoyant filter bioreactor, BFBR] under micro aerobic conditions. In the present attempt different media composition was tried to develop anaerobic protozoan cultures, which are important in understanding their specific roles in ecosystem and biodegradation. Among media,

Wheat extract in Chalkey's medium, Cellulose medium and Cellulose in Chalkey's medium, Supported the growth. Drastic changes in the pH level affected the growth of organisms, especially the cultures in wheat extract medium.

Introduction

Microbial diversity is an unseen national as well as international resource that deserves greater attention. Too small to be seen no longer means too small to be studied or valued. Microbial diversity encompasses the spectrum of microscopic organisms including bacteria, fungi, algae, and protozoa. These organisms populate the soil, water and air that surround us as well as live in more unusual environments such as the boiling water of hydrothermal vents, deep ocean trenches and alkali lakes.

Microbes: The earth's engine

Microorganisms have been evolving for nearly 4 billion years and are capable of exploiting a vast range of energy sources and thriving in almost every habitat. For 2 billion years microbes were the only form of life on Earth. During this long history, all of the basic biochemistries of life evolved, and all life forms have developed from these microbial ancestors. It is estimated that 50% of the living protoplasm on this planet is microbial. Microorganisms represent by far the richest repertoire of molecular and chemical diversity in nature. They underlie basic ecosystem processes such as the biogeochemical cycles and food chains, as well as maintain vital and often elegant relationships between themselves and higher organisms. Microbes provide the fundamental underpinning of all ecosystems. Without microorganisms, all life on Earth would cease.

Microbes: The biological frontier

Because microorganisms are small, they are least known, and this gap in knowledge is particularly apparent for bacteria and other small organisms. Current evidence suggests that perhaps 1.5 million species of fungi exist yet only 5% are described. For bacteria there may be 300,000 to 1 million species on earth yet only 3,100 bacteria are described in Bergey's Manual, the treatise of described bacteria. A gram of typical soil contains about 1 billion bacteria, but only 1% of those can be cultured. Similarly low fractions of microorganisms have been cultured from fresh water and ocean environments. Hence, most microbes remain to be discovered.

The value of microbial diversity

Focusing on microbial diversity is timely. Diverse microorganisms are essential to a sustainable biosphere. They are able to recycle nutrients, produce and consume gases that affect global climate, destroy pollutants, treat our wastes, and they can be used for biological control of plant and animal pests. The study of microbial diversity is also important to solve new and emerging disease problems and to advance biotechnology. New technologies, particularly in nucleic acid analysis, computer science, analytical chemistry, and habitat sampling and characterization place the study of microbial diversity on the cutting edge of science.

Humans over the ages have been highly successful in applying processes carried out by microorganisms to solve problems in agriculture, food production, human health, environmental quality, and industry. Recently developed technologies in molecular biology and genetics offer great promise for new opportunities to develop the potential of microbial diversity.

The Value of Microbial Biodiversity Research

- To expand the frontiers of knowledge about the strategies and limits of life, especially those thriving at extreme conditions and on unusual redox couples,
- Microorganisms are of critical importance to the sustainability of life on our planet,
- The untapped diversity of microorganisms is a key resource for new genes and organisms of value to biotechnology,
- Diversity patterns of microorganisms can be used for monitoring and predicting environmental change,
- Microbes play a role in conservation and restoration biology of higher organisms and degraded landscapes,

Microorganisms are the major sources of antimicrobial agents and produce a wide range of other important medicinal compounds including enzymes, enzyme inhibitors, antihelminthics, antitumor agents, insecticides, vitamins, immuno suppressants, and immuno modulators. These agents have all been discovered during the past 50 years and represent only a small portion of what is likely present in nature. Pharmaceuticals of microbial origin have a market value (at the wholesale level in the developed world) of approximately $35-50 billion annually.

Preserve microbial diversity

(a) Conduct research leading to the **preservation of mixed communities**, e.g., consortia, natural communities.

(b) Improve **culture preservation strategies**, and optimized regimes for difficult to preserve microbial groups.

(c) Maintain B and coordinated network of **culture collections** that are accessible worldwide.

(d) Conserve habitats with rare or threatened species.

Anaerobic waste water treatment

Application of anaerobic biotechnology for the domestic as well as industrial wastewaters leading to methanogenisis is now attracting worldwide attention. This is mainly because of the several advantages anaerobic biological process over aerobic treatment methods, which include low sludge production and nutrient requirements as well as production of useful byproduct in the form of methane.

Anaerobic conversion of organic matter is now recognized as a complex process involving coordinated activity of a number of different bacterial trophic groups .it involves consortium of microbes where obligate anaerobic bacteria form the dominant populations and is a multi stage biochemical process performed on different organic material.

No individual fermenting microbe has the ability to completely degrade complex organic compounds [eg. cellulose] to CO_2 and H_2 and this highlights a dramatic difference between microbial decomposition in the presence and in the absent of oxygen. In the absence of oxygen complete degradation to CO_2 requires the concerted activity of several different types of anaerobes, and even then some carbon may remain locked up in CH_4. A fundamental characteristic of the community of the microbial fermenters is the interdependence of different metabolic types, the products or activities of one organism provide substrate or an equable chemical environment, or both for neighboring microbes of different types. However, with in each of these types, many species may compete for the same resources, and two discernible strategies operate to enhance their relative selective advantage. The first is to maximize the efficiency and energy yield of catabolic pathways and thus in crease the potential growth rate. All fermenting organisms are forced to use pathways, which yield only relatively small amounts of energy, but specific metabolic innovations [eg. Releasing of H_2 gas] maximize the yield. The major draw back with an innovation such as H_2 evolution is that it almost always requires the close presence of a H_2 consuming organism to maintain the ambient H_2 pressure at a thermodynamically low level. These organisms use hydrogen as the source of energy.

$$4H_2+2CO_2 \longrightarrow CH_3COO^- + H^+ + 2H_2O + [G^0]\ -95\ KJ/mol$$

HOMOACETOGEN

$$4H_2+ CO_2 \longrightarrow CH_4+2H_2O+ [G^0]\ -131KJ/mol$$

METHONOGEN

[Fenchel and Finley 1995]

Thus maximizing energy yield may require coexistence with an unrelated organism having a complementary metabolism and selective advantage gained probably underpins the evolution of syntrophic associations.

Syntrophy

It can be broadly defined as metabolites of one type of organism serving as substrates for other types of organism. The anaerobic metabolism in natural communities is that, the catabolism of organic matter depends on the participation of several different types of organisms carrying out different functions.

One of the extra ordinary things about the diversity of anaerobic protozoa is that most have other microorganisms living either inside them or attached to their external surfaces. Those associations are usually symbiotic in so far as they are invariably permanent, they involve specific partners, and one or both partners benefit from the association. As group the anaerobic protozoa are remarkable for the high proportion of species, which live, in symbiotic association with prokaryotes. They are also unique in types of symbiotic partners they have acquired endosymbiotic purple non-sulphur photosynthetic bacteria and methanogens; and ectosymbiotic sulphate reducing bacteria. One obvious feature, which is common to this entire prokaryote group, is that they can consume hydrogen as a substrate and this point to the likely significance of hydrogen production by the protozoan in the maintenance of these symbiotic associations. One could argue that because of the efficiency of hydrogen evolving fermentation [and hence the fitness of the eukaryote performing it] will be increased by having a resident for hydrogen that evolution has favored the retention of this type of functional consortium [5].

The complete mineralization of organic matter by communities of anaerobic prokaryotes requires the integrated activities of several different forms [eg. cellulose degraders, VFA producers, sulphate reducers and methanogens] now with the introduction of anaerobic eukaryotes there is a quantum leap in the simplicity and efficiency of the whole decomposition process. The protozoan comes with the capacity to engulf the particles and with a suit of digestive enzymes and a metabolic repertoire superior to any thing found in the prokaryotes. The consequence is that the concerted activity of a variety of prokaryotes can be transferred to a eukaryote with single prokaryotic symbiont species and complex organic matter ingested by the eukaryote ends up as CO_2, plus either H_2S or CH_4 [5].

Culturing of protozoans

Unlike the bacteria, protozoa are difficult to cultivate in vitro for extended periods. There for, most research has emphasized the bacterial community. The role of biochemical capabilities of protozoa is less fully under stood [18]

Metabolic studies on free-living anaerobic protozoa have been hampered by the limited number of organisms growing in axenic cultures [6] and culturing of anaerobic protozoans reported in limited number of research papers [6,7].

It have been reported that syntrophic relation between anaerobic protozoans and anaerobic bacteria accelerating degradation rates when compared with prokaryotic consortium [5]. But their significance in wastewater treatment plants is less fully under stood. Hence culturing of anaerobic protozoans became significant, as it is the preliminary step for studying their role in biodegradation and biochemical pathways in wastewater treatment.

Hence the objective of the work was to develop suitable culture techniques for the growth of protozoans isolated from anaerobic reactor sludge [buoyant filter bioreactor, BFBR] under micro aerobic conditions. In the present attempt different media composition was tried to develop anaerobic protozoan cultures, which are important in understanding their specific roles in ecosystem and biodegradation.

Materials and Methods

Composition of media

Wheat extract medium

50 ml wheat extract is prepared by boiling 5 gm wheat grains in Chalkey's medium **[Chalkey's medium** can be used instead of

distilled water as a more refined media. This consists of 0.1 gm NaCl, 0.004 gm KCl and 0.006 gm CaCl in 1000 ml.].

Cellulose media

1. 5 gm crystalline cellulose and 20 ml phosphate buffer are added to 500 ml distilled water
2. 5 gm crystalline cellulose and 20ml phosphate buffer are added to 500 ml of Chalkey's medium.

A nutrient solution is prepared using a combination given below

$NaHCO_3$	400 mg/L
NH_4Cl	280mg/L
KH_2PO_4	250mg/L
$MgSO_4.7H_2O$	100mg/L
$CaCl_2.2H_2O$	10 mg/L

2 ml nutrient solution is added to both cellulose media

Resazurin [4ml] was added to cellulose medium as indicator show contamination of O_2.

Experimental set up

- Different compositions used for culture protozoans are,
- Wheat extract in Chalkey's medium,
- Cellulose medium and
- Cellulose in Chalkey's medium supported the growth.

Anaerobic culture system was tried in 50ml and 500ml bottles. Wheat extract medium is prepared in 50ml bottles [SCHOOT DURAN, GL 32, amber coloured, made in Germany] having screw cap with aperture to which a rubber septum can be attached. Through the rubber septum a syringe was inserted to collect samples or to add buffer and for inoculation. Cellulose medium was prepared in 500ml bottles [SCHOOT DURAN, colorless, made in Germany]. The bottles were closed with a rubber cork, having two outlets. The system covered with paraffin sheath in order to prevent gas exchange [Shown in the photograph].

Results and discussion

In the present attempt different media composition was tried to develop anaerobic protozoan cultures, which are important in understanding their specific roles in ecosystem and biodegradation.

As the growth curve indicates slow growth rate in anaerobic systems when compared to aerobic systems. Drastic changes in the pH level affected the growth of organisms, especially the cultures in wheat extract medium. Adjustment of pH by adding sodium bi carbonate solution [1N] was needed to maintain the growth. Desirable pH was found as 6.8 to 7.2.

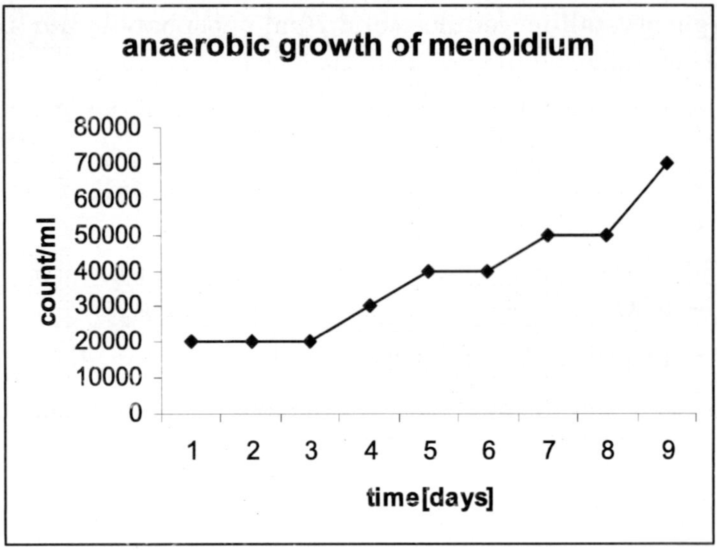

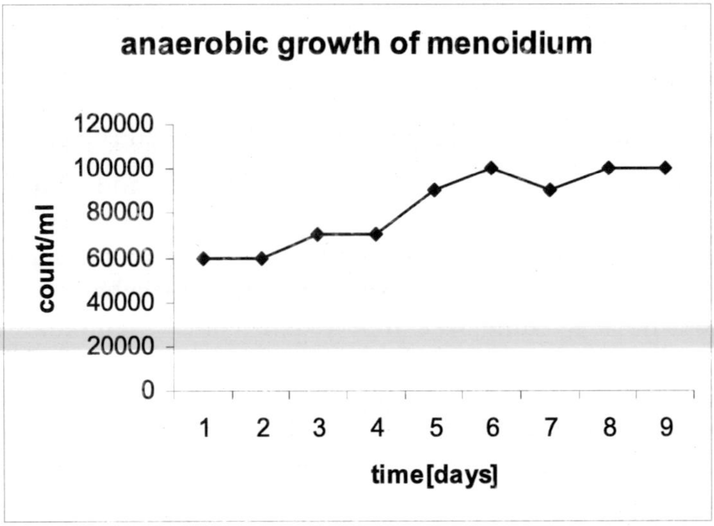

2-3 days was needed to have exponential growth in the wheat extract medium, but in the cellulose medium 2 weeks was needed to observe the growth of *Menoidium*. This might be due to the osmotrophic nutrition of *Menoidium*. Persumably as the cellulose medium did not contain any soluble carbon source, a period of two weeks might be needed to degrade cellulose in to soluble form [glucose] by bacterial degradation that could be consumed by *Menoidium*. Presence of glucose in the cellulose media was observed during the study.

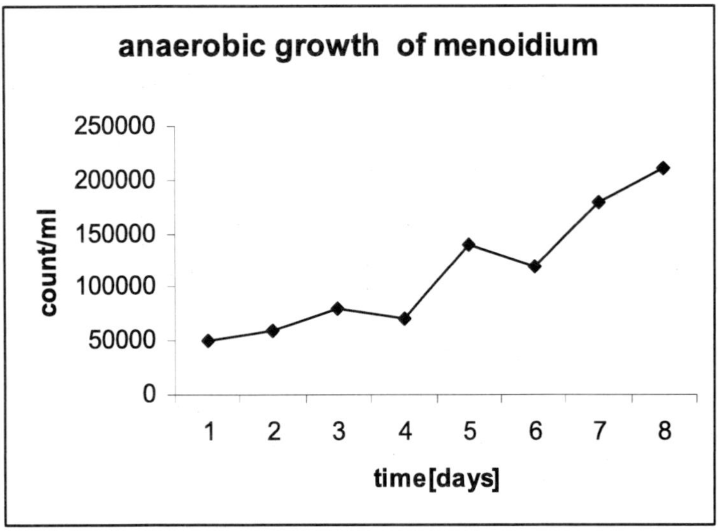

References

Christopher F. Foster and D.A. John Wase 1989. E*nvironmental biotechnology*, Ellis Horwood limited

Curds C.R 1969. An illustrated key to the British fresh water ciliated protozoa commonly found in activated sludge [Water Poll. Res. Tech Paper no. 12 . Her Majesty's Stationary Office London, England].

Downing and Wheatland 1962. fundamental considerations in biological treatment. transactions of the institute of chemical engineers 40, 91-103

Fenchel, T and Finley, B.J. 1990. Anaerobic free-living protozoa: Growth efficiencies and the structure of anaerobic communities. FEMS Microbiology. Ecology 74, 269-276

Fenchel, T and Finley, B.J. 1995. *Ecology and evolution in anoxic worlds.* Oxford University Press. Oxford.

Giancarlo A. Biagini *et al.* 1998. Carbohydrate and amino acid fermentation in the free-*living* primitive protozoan Hexamta sp. *Applied environmental microbiology* vol. 64.

Giancarlo A. Biagini, B.J. Finlay and David Lloyd 1998. Protozoan stimulation of anaerobic microbial activity: enhancement of the rate of terminal decomp-osition of organic matter. FEMS Microbiology Ecology. pp. 1-8

Giancarlo A. Biagini and Catherine Bernard 2000. Primitive anaerobic protozoa. Microbiology comment, Microbiology 146, pp. 1019-1020.

Martin and Muller 1998. The hydrogen hypothesis for the first eukaryote *Nature* 392, 37-41.

Muller, M. 1988. Energy metabolism of protozoa without mitochondria. Annual review microbiology, 42,465-488.

Muller 1991. Energy metabolism of anaerobic parasitic protests International biochemical protozology.

Patterson, D.J 1996. Free living fresh water protozoa. Manson publishing Ltd. London.

Schofield 1971. *some biological aspects of activated sludge plant at Leicester,* water pollution control 70, 32-47.

Sreenivas Kota *et al.* 1999. Influence of protozoa grazing on contaminant biodegradation. FEMS Microbiology Ecology 29. 179-189.

Standard methods, 19th edition 1995. American Public Health Association [APHA]

Sydenham, 1971. *A reassessment of the relative importance of the ciliates in the ecology of activated sludge,* Hydrobiologia 38.

Williams, A.G. and Coleman, G.S. 1991. The rumen protozoa. Springer Verlag, Berlin.

Williams, A.G. 1986. *Rumen holorich ciliate protozoa,* microbiological reviews, pp. 25-49.

DIVERSITY OF AM FUNGI IN SOME CONTAMINATED SITES OF KERALA

M.V. Bindu and V.S. Harikumar

Department of Post Graduate Studies and Research in Botany, Sanatana Dharma College, Alappuzha -688 003, Kerala

Abstract

A study was conducted to assess the diversity of arbuscular mycorrhizal (AM) fungi in some polluted sites of Kerala. The study revealed wide variation in AM population and diversity in the sites studied. Maximum spore load was observed in soils polluted with organic wastes. The soils polluted with industrial wastes had a low level of spore population. Various AM fungal species isolated in the contaminated sites belonged to the genera *Acaulospora, Entrophospora, Glomus, Sclerocysists* and *Scutellispora. Glomus* species was encountered in all the sites.

Introduction

The rapid growth of various industries in recent years is creating more waste lands through the effluent spread, which contains toxic elements and heavy metals. While some of them form plant nutrients, most of them are phytotoxic, hampering the establishment of a plant cover on such wastelands. The potential of Arbuscular mycorrhizal (AM) fungi is being harnessed in afforestation and rehabilitation of wasteland (Hetric *et al.,* 1994). Mycorrhizal symbiosis offsets the toxicity by way of acting as a biofitter for pollutents. Only a few attempts have been made to ascertain the impact of physiochemical factors on the composition of AM flora. The present study reports diversity of AM species in five polluted soils of Kerala.

Materials and Methods

Five polluted soils of Alleppey and Kottayam district of Kerala during the three seasons are selected for the study

Ten replicate soil samples were withdrawn from each polluted location from a depth of 10 cm. The soils were kept in polythene bags, labeled and stored at 4°C till analysis. AM fungal species were recovered from the soil samples by the wet-sieving and decanting method (Gerdemann and Nicolson, 1963). The identification of the AM species was done as per the manual of Schenck and Perez (1987). The physio-chemical characteristics of polluted soils were analysed as per standard procedure (Byju, 2001).

Results

Wide variation existed in the physicochemical characteristics of the soils understudy (Table 1). The soils of site 1 and 4 showed acidic tendency, while in other sites the soils were alkaline. The soil moisture was high in all sites except in site 3. Irrespective of sites, the soils had high levels of Organic Carbon. Higher values for nitrogen (N), phosphorus (P) and potassium (P) was observed in site 5.

Table 1. Physicochemical chacteristics of polluted soil (mean ± SD, n = 10)

Sl. No.	Study site	Soil type	Temperature (°C)	pH	Moisture (%)	Organic Carbon (%)	N (kg/ha)	P (kg/ha)	K (kg/ha)
1.	Agricultural	Black	27.31± 1.38	4.42± 0.51	12.29± 9.53	1.19± 9.53	1.19± 104.40	431.9± 104.40	24.91± 20.12
2.	Cement part	Laterite	29.12± 1.05	7.71± 0.65	9.86 ± 3.76	1.27± 0.44	520± 91.32	63.28± 56.28	86.48± 56.88
3.	Seafood effluent	Sandy	27.50± 1.00	7.69± 0.51	8.31± 2.63	1.28± 0.36	425.6± 86.35	315.90± 111.33	109.40± 92.45
4.	Coir retting effluent	Sandy	27.80± 0.40	5.55± 1.28	21.56± 5.02	1.03± 0.16	378± 83.06	149.90± 118.62	76± 55.11
5.	Municipal solid waste	Sandy	28.10± 0.53	7.72± 0.47	9.75± 1.64	1.43± 0.56	450.8± 134.95	309.30± 134.95	129.20± 122.19

Spore population of AM fungi in soils differed with the sites and seasons can alter the distribution and composition of AM fungi. It was shows the ability of AM fungi. It also shows the ability of AM fungi sustain in such a stressful ecological situations and the possible use of AM fungi to sustain in such a stressful ecological situations and the possible use of AM fungi as bioindicutors of pollon or as bioremediation agents in polluted soils.

Table 2. Spore load and AM species diversity in polluted soil

Sl. No.	Study Site	Spore Count /50 ml soil	AM Species
1.	Paddy field	7.50 ± 3.00	*Entrophospora* sp. *Glomus* sp. *Sclerocysists* sp.
2.	Cement dust polluted	3.50 ± 2.33	*Acaulospora* sp. *Thromi, Glomus desertticola, glomus fasciculatom* & *Glomus* sp.
3.	Sea food effluent	3.50 ± 2.33	*Glomus* sp.
4.	Coir retting effluent	6.7 ± 2.36	*Glomus australe, Glomus fragilostratum, Glomus* sp. *Scuetellispora perisca*
5.	Organic waste polluted	10.83 ± 4.49	*Acaulospora delicate, Glomus microaggregatum, Glomus* sp., *Scutellispora perisco*

Conclusion

The study reveals that various soil pollutants and seasons can alter the distribution and composition of AM fungi. It also ahows the ability of AM fungi to sustain in such a stressful ecological situations and the possible use of AM fungi as bioindicators of pollon or as bioremediation agents in polluted soils.

References

Byju, G. 2001. Soil Analysis: A Laboratory Manual, CTCRI Sreekariyam Thriuvanathupuram.

Gerdemann, J.W. and Nicolson, T.H. 1963. Spores of mycorrhizal Endogone species extracted from soil by wet sieving and decanting. *Trans. Br. Mycol. Soc.*46:235.

Hertrick, B.A.D., Wilson, G.W.T. and Figge, D.A.H. 1994. The influence of mychorrizal symbiosis and fertilizer amendments on establishment of vegetation in heavy metal mine spol. *Enviorn. Pollut.* 86:171-179.

Schenck, N.C. and Perez, Y. 1987. *Manual for the identification of VA mycorrhizal fungi*, University of Florida, Gainesville USA.

SCOPE OF INDEGENOUS RICE CULTIVARS FOR CONSERVING BIODIVERSITY

Sverup John, D. Alexander, Bindu, M.R. and P. Sushamakumari

Onattukara Regional Agricultural Research Station,
Kerala Agricultural University, Kayamkulam - 690 502.

Introduction

The staple food of Keralite is rice *(Oryza sativa)*. The average per capita consumption of rice is 70 kg per year. But 70 percent of the rice requirement is met from other States. Hence considering the food security, we have to give prime importance for the cultivation of rice in our own locality itself. More over, the rice cultivation has got a great role in ecosystem by conserving biodiversity and water resource of the region.

In Kerala, rice is cultivated in an area of 3.5 lakh hectare with grain production of 7.5 lakh tones annually (Anon., 2003). The important rice ecosystems of the state are Laterite Midland, Malayoram, Palakkad plains, Kuttanad, Black soil (Chittoor), Onattukara, High ranges, Kole and Pokkali. The corresponding share in area of each rice ecosystem of the state is presented in Table 1.

Since rice is an ancient food crop of Kerala, each ecosystem possesses indigenous cultivars. These genotypes fit well into the specific ecosystem without hampering the biodiversity.

Table 1. Important rice ecosystem in Kerala

Ecosystem	Percentage
Laterite midland	46.89
Malayoram	18.14
Palakkad plains	10.60
Kuttanad	6.70
Black Soil (Chittoor)	6.51
Onattukara	5.45
High ranges	4.83
Pokkali	0.88

(Sasidharan *et al.*, 2002)

Presently the high yielding variety coverage in area during virippu is 65 percent, mundakan is 55 per cent and summer is 96 per cent (Anon., 2003). The popularization of high yielding varieties has resulted in the loss of the precious indigenous cultivars. This can be termed as 'genetic erosion'. The indigenous cultivars, even though not high yielding, but possess good quality and resistance to biotic and abiotic stress situations (Vaughan, 1994). The details of indigenous cultivars pertaining to Onattukara ecosystem are presented in Table 2.

Table 2. Description of indigenous rice cultivars

Land races	Duration (days)	Height (cm)	Panicle length (cm)	No.of Grains/ panicle	100 grain wt. (g)	Remarks
Virippu						
Karavala	90	76	18	93	2.79	Tolerant to drought
Kochuvithu	88	130	18	110	2.32	Good cooking quality
Aramvella	128	135	20	60	1.79	Uniform flowering
Padanavella	129	111	16	60	1.83	Awnless, high yield potential
Edakkedavella	124	115	20	55	2.49	
Thavalakkannan	118	125	22	116	2.21	White and red types seen

Mundakan						
Chempavu	150	150	25	100	2.40	Good grain quality & high straw yield
Mundakan	138	130	25	121	2.97	Good quality
Athikkirazhy	128	122	23	99	2.78	Tolerant to biotic stress
Cherady	165	140	23	87	2.93	Suited to eastern lateritic regions
Mallikkuruva	125	116	17	63	2.55	Bold grains
Kilikkuruva	140	135	22	93	1.97	Small grains
Vayalithoova	165	175	25	60	3.05	Plant grows up according to water level
Specialty rice						
Njavara	115	105	18	84	2.49	Medicinal

(1) Virippu season

The *virippu* season represent the first crop (April-May to September-October) during which, short duration and photo-insensitive varieties are grown. Usually these varieties are dry sown during the onset of monsoon. The virippu land races are: *Karavala, Kochuvithu, Aram vella, Padanavella, Edakkeda vella* and *Thavalakkannan*. All these possess distinguish characters. The variety *Karawala* offers resistance to drought and *Padanavella* offers aspects for high yield. In Onattukara *Kochuvithu* was the traditional virippu variety grown for its early duration and adaptability. But since the introduction of high yielding varieties, *kochuvithu* is not prevalent at present

(2) Mundakan season

The Mundakan varieties represent the second crop (September-October to December-January) during which photosensitive varieties are grown. These long duration varieties are transplanted. The Mundakan land races are: *Chempavu, Mundakan, Cherady, Mallikkuruva, Kilikkuruva, and Vayalithoova.*

These genotypes possess varying degree of tolerance to major pests. They are invariably high straw yielding. The *Cherady* varieties are suited specific to the eastern latereitic region (John and Alexander 2001). *Vayalithoova* grows tall according to rise in level of water. It can be utilized for flood prone area and rice-fish culture programme. The inherent defect of this variety is that it sheds the grains in the field on maturity.

(3) Koottumundakan

Koottumundakan is a special system of cultivation prevalent in the rainfed lowlands of Palakkad, Alappuzha and Kannur Districts of Kerala, in which a mixture of seeds of photo insensitive (virippu) variety and photosensitive (mundakan) variety are mixed in a definite proportion 70: 30 (w/w) and sown together during the kharif (virippu) season. In Alappuzha district Koottumundakan is presently being practiced in Panavalli, Cherthala, Kuthiathode, Aryadu and Mararikulam. Here it is presently covering an area of 2000 ha. This system is practiced in areas were sowing/planting of Mundakan crop is not possible due to excess water in the field. A mixture of two varieties are sown in the field during first crop season will be harvested during August- September and second crop variety which is photosensitive, will be harvested during December-January. Traditional tall *indica* rice varieties, which are low yielding, are currently being used for this type of cultivation and the returns from the crop is rather low to the tune of around 2 t/ha. The farmer's combination is Chenkayam + Chitteni (Leena Kumari *et al.,* 2002)

(4) Specialty rice (Medicinal)

Njavara is the indigenous medicinal cereal plant of Kerala. It is widely used in the Ayurvedic system of medicine, especially in Panchakarma treatment. For aurvedic treatment the grains of Njavara variety is the main component. Along with njavara grains the medicinal plants viz: *Kurunthotty, Amrit, Adalodakom, Kanda karichunda, Asokam* and *Koduveli* are the other ingredient for *njavarakizhi*. There are two types of Njavara-black and white, based on glume colour. This unique rice having high content of free amino acids also possess the medicinal properties to rectify the basic ills affecting the circulatory, respiratory and digestive systems (Anon 2004). The main problem faced by Njavara growers of the region is the low setting of grain. Being medicinally valued, Njavara grains are getting premium price in the market.

(5) Variability of land races

Since the land races are seen in the isolated tracts, there is possibility of diversity and variability of biometric characters. In the case of Cherady, the photosensitive mundakan variety, during the survey 144 types could be located. The variability expressed for different morphological characters are presented in Table 3.

Table 3. Variability analysis of local Cherady.

Characters	Range	Mean
Plant ht. (cm)	103-176	130
Tiller No.	8-16	13
Panicle length (cm)	19-30	23
No. grains /panicle	76-184	87
1000 seed wt (g)	23.7-32.3	29.3
Seed yield (g/$M^{2)}$	123-608	344
Straw yield (g/$M^{2)}$	150-1062	854

This indicates the existence of variability for almost all the morphological characters retaining the key characters of the genotype (John & Bindu 2002).

Conclusion

Since rice is an ancient food crop of Kerala, each rice ecosystem possesses indigenous cultivars. The indigenous cultivars pertaining to Onattukara rice ecosystem is discussed here. The indigenous virippu rice cultivars are *Karavala, Kochuvithu, Aram vella, Padanavella, Edakkeda vella* and *Thavalakkannan*. The indigenous Mundakan rice cultivars are *Chempavu, Mundakan, Athikkirazhy, Cherady, Mallikuruva, Kilikuruva* and *Vayalithoova*. Under the specialty rice group, *Njavara* the medicinal rice deserves much scope for popularization. Koottumundakan is a traditional rice cultivation practiced in certain pockets of the region. These land races possess varying degree of variability for different traits, which offers scope for genetic improvement by breeding methods. Hence collection, conservation, characterization and documentation of all the available land races, which are facing extinction, should be taken up.

References

Anon, 2003. Farm Guide 2003, Farm Information Bureau, Govt. of Kerala.

Anon, 2004. The Hindu ' *Njavara* facing extinction' dt 7th June 2004.

Leenakumari, S., Rose Mary Francis, G. Valarmathi and Gregory Zackariah 2002. National Symposium on Priorities and Strategies for Rice in High Rainfall Tropics, Kerala Agrl. University, 10-11 October, 2002 - page 11.

John, Sverup and D. Alexander 2001. 11th Swadeshi Science Congress, Kerala Forest Research Institute, Peechi, 7-9 November, 2001- page 32.

John, Sverup and M.R. Bindu 2002. National Symposium on Priorities and Strategies for Rice in High Rainfall Tropics, Kerala Agrl University, 10-11 October, 2002, page 3.

Sasidharan, N.K, K.C Rajan and P.V. Balachandran 2002. National Symposium on Priorities and Strategies for Rice in High Rainfall Tropics, Kerala Agrl University, 10-11 October, 2002- page 90-103

Vaughan, D.A. 1994. The Wild Relatives of Rice: A Genetic Resources Handbook. International Rice Research Institute, Manila, Philippines.

WILD SESAME SPECIES DISTRIBUTION IN SOUTHERN KERALA

Sverup John, Manoj Sebastian, D. Alexander, Shyam. S. Kurup and K. Rajmohan

Kerala Agricultural University,
Onattukara Regional Agricultural Research Station, Kayamkulam-690 502

Introduction

Sesame (*S. indicum* L.) is one of the ancient annual oilseed crop cultivated in Kerala, where the genus exhibited greater diversity. It belongs to family *Pedaliaceae.* In sesame, wild species are potential source of many desirable traits such as resistance to pests,disease and drought. An extensive survey was conducted for identifying wild sesame in six southern districts of Kerala *viz.* Thiruvananthapuram, Kollam, Alappuzha, Pathanamthitta, Kottayam and Ernakulam. Species diversification has been observed at the time survey.

Result and discussion

Based on the survey three species of wild sesame were identified viz; *Sesamum malabaricum, Sesamum mulayanum* and *Sesamum radiatum*. All these species are seen near road side and waste lands. The wild species *S. malabaricum* and *S. mulayanum* have the same chromosome number (2n=26) that of cultivated *S. indicum,* where as the wild species *S. radiatum* has 2n = 64. *S. malabaricum,* growing near the road side and waste land in Kerala has been recorded (Desai and Goyal, 1981). In Index Kewensis (Joshi, 1961) there is a record of *S. malabaricum* from Kerala. The wild species S. *mulayanum* (Nair, 1963) and *S. radiatum* (Mathew, 1999) reported from waste land of Kerala. There is a new record of *S. radiatum* from Thiruvananthapuram district, Kerala (Augustine, 2001)

In the present study the population percentage of *S. malabaricum* is 4.99, *S. mulayanum* is 9.01 and *S. radiatum* is 85.99. Among the three species, *S. radiatum* distributed widely in all districts surveyed (Table 1). Based on the survey conducted in six districts of southern Kerala, the maximum diversity of wild species are seen in Alappuzha district.

Table 1. Distribution of wild species

Districts	Locality	Species
Kottayam	Changanassery	■
	Kuruchy	■ #
	Kottayam	■
Ernakulam	Aluva	■
Pathanamthitta	Thiruvalla	■#
	Pandalam	■#
	Adoor	■
Alappuzha	Chenganoor	#*
	Cherthala	■
	Haripad	■#*
	Thottappally	■#*
	Punnappara	■#*
	Ambalapuzha	■#
Kollam	Kollam	■
	Kottarakkara	■
	Nendakara	■#*
	Karunagappally	■#*
Thiruvananthapuram	Kilimanoor	■
	Vellayani	■

*S. malabaricum** *S. mulayanum* # *S. radiatum* ■

Discription of wild species

(a) *S. malabaricum* an erect herb, 146-187 cm height, leaves opposite, trifoliate and toothed below, 6.4 to 20.5 cm length and 4.5 cm to 6.5 cm breadth. Upper leaves alternate, narrowly elliptic to lanceolate. Flower axillary, solitary, shortly pedicelled, purple colour with deep pink spot inside the corolla. Bracts 2, green, having two globular yellow glands in the axil of the bracts; calyx green hairy sepals 5, corolla oblique, hairy, glabrous within, lobes 5, 4 equal lips and one longer lip, stamens 4 didynamous, connective surrounded by a yellow gland at the apex,anther cell oblong, suture black in colour, ovary hairy stigma lobed, spatulate, capsule erect, oblong, 2.4 x 0.8 cm, seed obovate, winged, reticulate and black. 1000 seed weight 2.3 gm and seed oil content is 24 percent. Duration of plant 125-130 days.

(b) *S. mulayanum* is an erect herb, 60-90 cm height, leaves opposite, trifoliate and toothed below, 9.2 to 15.6 cm length and 3.2 cm to 6 cm breadth. Upper leaves alternate, narrowly elliptic to lanceolate. Flower axillary, solitary, shortly pedicelled,purple colour with deep pink spot inside the corolla Bracts 2, green, having two globular yellow glands in the axil of the bracts. Calyx green hairy sepals 5 corolla oblique, hairy, glabrous within, lobes 5, 4 equal lips and one longer lip, stamens 4 didynamous connective surrounded by a yellow gland at the apex, anther cell oblong, suture black in colour, ovary hairy stigma lobed, spatulate, capsule erect, oblong, 2.1 x 0.98 cm, seed obovate, winged, reticulate and black in colour. 1000 seed weight 2.26 gm. and 32 per cent seed oil content 110-120 days duration.

(c) *S. radiatum* is an erect herb, annual but shows perennial, continuous branching from the base of the stem contribute perennial habit. 145-190 cm height leaves simple, opposite below alternate above the lower leaves ovate and toothed below 12.5 cm to 17.6 cm length and 2.7 to 5.6 cm breadth, upper leaves elliptic to lanceolate. Flower axillary, pinkish violet flowers with spotted purple within. Bract 2 having globular deep purple glands in the axil of the bracts, calyx green, lobes 5, hairy. Corolla companulate, oblique, pinkish-violet to whitish-rose within and spotted within, petals 5 four equal and Stamen 4, didynamous, anther oblong, connective surrounded by a purple gland at apex. Ovary dense hairy, stigma 2 lobed, spathulate. Capule erect, oblong, 4 angled and grooved. dense pubescent, splitting apex down ward, seeds many, ovoid, 3 x 2 mm black, brown with transverse ridges, emanating from a central field on the broad surfaces especially the flatter one. 1000 seed weight 2.006 gm. more than six months duration and with 33 per cent seed oil content.

The three wild species distributed in southern Kerala are *S. malabaricum, S. mulayanum* and *S. radiatum*. These wild species showed seed dormancy due to hard seed coat. The wild sesame species viz; *S. malabaricum, S. mulayanum* have the same chromosome number of cultivated sesame (2n = 26) and possess resistance to majority of biotic and abiotic stress condition, hence these species can be utilized for resistance breeding resorting to back cross programme. Since the chromosome number of *S. radiatum* is 2n = 64, this isolation barrier limits the transfer of resistance gene from this wild species by conventional breeding programme.

References

Augustine, K.T. 2001. New record of *Sesamum radiatum* Schumach & Thonn *(Pedaliaceae)* from Thiruvananthapuram district, Kerala. *J. Econ. Taxon. Bot.* 25(3): 617-620.

Desai, N.D. and Goyal S.N. 1981. Integrating breeding objectives and agricultural practices for Indian conditions, FAO, Plant production and protection 29:118-120.

Joshi, A.B. 1961. Sesamum Indian Central oilseeds committee, Hydrabad Ist Ed. pp. 109.

Mathew K.M. 1996. *Illustration on the Flora of the Palani Hills, Southern India*. The Rapinat Herbarium, Tiruchirapalli. 557.

Mohanan, M. and Henry, A.N. 1994. *Flora of Thiruvananthapuram*, Kerala, B.S.I., Calcutta. pp. 339-340.

BIODIVERSITY OF THE SOIL FUNGI IN THE SEMI-ARID TROPICAL SOILS OF PUTTA-PARTHY, VIDYA GIRI, PRASANTHINILA - YAM CAMPUS, ANANTAPUR DISTRICT, A.P.

B.S. Vijayakumar, Sathya Ganapathy Bhat, Vivek Pahwa, K. Premananda, K.N. Naresh, K. Krishna Bhat, and Sanjeev

Department of Biosciences, Sri Sathya Sai Institute of Higher Learning, Prasanthinilayam, Ananthapur district, A.P

Abstract

During the present investigation on the Biodiversity of soil micro-fungi in the semi-arid soils of Vidyagiri campus in Puttaparthy, Anantapur district, A.P., the dilution plate technique was used to isolate soil fungi. The soil fungi isolated and identified were: *Aspergillus niger, A. flavus, A. nidulans, A. ochraceus, A. terreus, A. flavipes, A. parasiticus, A. ustus, A. fumigatus, Penicillium steckii, P. islandicum, P. rubrum, Fusarium solani, F. oxysporum, F. dimerum, Rhizopus nigricans, Chaetomium thermophile, Neocosmo-spora vasinfecta and Trichoderma viride.* During the period of present investigation (2003-2004) *Trichoderma viride* was dominant and in frequent occurrence and this is interesting because it can be used as a Biocontrol agent to control pathogenic fungi.

Introduction

Biodiversity of soil microfungi are important in the field of soil microbiology and a good deal of work has been carried out for the last few decades all over the world. However no attempt has been made so far to understand the biodiversity of soil microfungi in the semi-arid tropical soils of Puttaparthy, A.P. Therefore, an attempt has been made to understand the biodiversity of soil fungi in the semi-arid soils of Putaparthy, Pasanthinilayam, Campus, Anantapur district, A.P.

Materials and Methods

Soil samples were taken in and around from semi-arid regions of Puttaparthy, A.P. Soil fungi were estimated as per the procedures given by Trivedi and Goel, 1986. Soil fungi was isolated and identified as per the procedures of Manoharachary, 2000.

Results and Discussion

During the present investigation on the Biodiversity of soil microfungi in the semi-arid soils of Vidyagiri campus in Puttaparthy , Anantapur district, A.P., the dilution plate technique was used to isolate soil fungi. The soil fungi isolated and identified were: *Aspergillus niger, A. flavus, A. nidulans, A. ochraceus,A. terreus, A. flavipes, A. parasiticus, A. ustus, A. fumigatus, Penicillium steckii, P. islandicum, P. rubrum, Fusarium solani, F. oxysporum, F. dimerum, Rhizopus nigricans, Chaetomium thermophile, Neocosmospora vasinfecta* and *Trichoderma viride.* During the period of present investigation (2003-2004) *Trichoderma viride* was dominant and in frequent occurrence and this is interesting because it can be used as a Biocontrol agent to control pathogenic fungi. Joffe and Borut (1966) found species of Aspergillus, Penicillium, Fusarium, Rhizopus, Alternaria, Sclerotium and *Rhizoctonia, Trichoderma as* predominant fungi. Among these Trichoderma viride was the predominant fungi in the fields of Israel. The very same fungus was found in the present investigation to be associated continuously in the semi-arid soils of Puttaparthy, Anantapur district, A.P. Altogether 45 species of fungi were isolated and identified from the semi-arid soils of Puttaparthy, Anantapur district, A.P. during the period of the year (2004). Aspergillus, Penicilium, and Fusarium species were also predominant and frequent in occurrence

Acknowledgement

The authors express their deep sense of gratitude to **Bhagawan Sri Sathya Sai Baba** for His grace and guidance during the period of the study. The authors are thankful to UGC for providing funds in the form of sanctioning **UGC Major research project**. The authors are thankful to the authorities of SSSIHL, Prasanthinilayam, for providing the facilities to carry out this work in the Institute.

References

Joffe, A.Z. and S.Y. Borut 1986. Soil and kernel mycoflora of groundnut fields in Israel. *Mycologia*, 58: 629-640.

Manoharachary, 2000. Microbial ecology of soil fungi., *Nat. Acad. Sci. Letters*. 4(8):305-309.

Trivedi, R.K. and P.K. Goel 1986. *Chemical and biological methods for pollution studies*. Environmental Publications, Karad, India.

INTRA-SPECIFIC VARIATIONS OF TRICHOMES IN THE GENUS *SOLANUM* L.

Anjana, V.N., N. Omana Kumari and P.M. Mathew

Department of Botany, University of Kerala, Kariavattom

Abstract

The trichome characteristics of eleven species belonging to the genus *Solanum* L. were studied. The genus presented a wide range of variation with many of the members exhibiting up to 5-6 different types of trichomes, while a few like *Solanum wendlandii* are glabrous. Maximum diversity in the trichome types was observed in *Solanum torvum, Solanum nigrum, Solanum melongena, Solanum pubescence, Solanum indicum* and *Solanum macranthum. Solanum nigrum* with herbaceous habit and very small flowers was observed to be the most advanced with bicellular and multicellular finger hairs, multicellular glands, medium sized multi-radiate stellate hairs and unique multicellular finger hairs with constricted cells and very narrow tapering apical cells. *Solanum aculeatissium* with primitive shrubby habit and medium sized flowers, possessed glandular or finger trichomes, which were unbranched. The variations in the trichome features could thus be correlated with the evolutionary trends from the morphological stand point of view also. The study highlights the significance of trichomes as stable tools in the evaluation of existing systematic classifications and may provide supplementary evidence in tackling taxonomic problems.

Introduction

The genus *Solanum* Linn. (Solanaceae) with over 700 species (Hunziker, 2001) distributed in the warmer regions of the world, is represented by over 38 species in India. The genus has been known to exhibit the greatest diversity with regard to the trichomes within the family. Although several taxonomists have studied the trichomes of the family in general (Metcalfe and Chalk 1950; Ogg *et al.* 1981; Lavania 1990) no attempt has so far been made to study the diversity of the trichomes within the genus *Solanum* from South India. The present study is aimed at filling this lacuna.

Materials and Methods

Trichomes of eleven species (Table) of the genus *Solanum*, collected from different localities of South India were studied, from epidermal peelings taken following the methodology standardized by Ahmed (1964). The trichomes were classified following the terminology proposed by Seithe (1979).

Observation

Solanum nigrum L.

Annual with glabrous, toothed ovate leaves and small white flowers in cymes; berries globose, purplish black and valued medicinally.

Multicellular, broad finger trichomes, bi- or multicellular narrow finger trichomes and multicellular glands were observed on both the surfaces of the leaf (Table). Lower epidermis with multi-radiate stellate hairs.

Solanum spirale Roxb.

A branched shrub with glabrous stem. Leaves elliptic and entire; long pedicels bearing small white flowers. Berries glabrous, smooth, and yellow.

The upper epidermis glabrous; lower epidermis with finger trichomes and mlticellular glands.

Solanum torvum Swartz.

A shrub armed with prickly stem; leaves ovate and tomentose with small prickles, flowers white in denser cymes; berry globose and green.

Upper epidermis with finger trichomes, multicellular gland trichomes and stellate trichomes - tri-, tetra- or multi-radiate. The lower epidermis showed finger trichomes, multicellular glands and tri-multiradiate stellate trichomes.

Solanum indicum L.

A branched undershrub, large prickles, sharp, often slightly recurved; stem purple and covered with hairs; leaves prickly on both sides and densely clothed with hairs; flowers purple blue in cymes; berry globose and dark yellow when ripe.

Unicellular finger, multicellular gland tipped and stalked multi-radiate stellate hairs on the upper epidermis. The lower epidermis showed bicellular finger hairs and tetra- radiate stellate hairs also.

Solanum pubescens Willd.

A tomentose shurb; leaves ovate, slightly cordate at base; flower cymes racemose; corolla purple. Berry red, glabrous, globose.

The upper epidermis was more diverse with multicellular gland tipped, unicellular finger, gland tipped finger and multi-radiate stellate hairs. The lower epidermis showed both multi-radiate stellate hairs as well as gland tipped multi-radiate stellate hairs with multicellular stalks.

Solanum melongena L.

An erect herb covered by soft hairs; leaves elliptic; flowers blue, segregated in to cymes; berry glabrous, dark purple, commonly referred as the Egg plant or the Brinjal.

Upper epidermis with unicellular finger trichomes and tri- and multi- radiate stellate hairs. Lower epidermis with uni-cellular finger trichomes, multicellular glands and stalked multi-radiate stellate trichomes.

Solanum trilobatum L.

A climbing undershrub; stem with recurved prickles; leaves ovate; corolla purple coloured; berry globose, scarlet.

Upper epidermis glabrous and lower epidermis with medium sized multi-radiate stellate hairs.

Solanum macranthum Dun.

A shrubby tree, commonly called the potato tree; leaves elliptic with serrate margins; flowers in shades of blue, purple and white in cymes; fruits small and oval.

Unicellular and bicellular finger and multi-radiate stellate trichomes were observed in the upper epidermis and multicellular glands and tri-, tetra- and multi-radiate stellate trichomes in the lower epidermis.

Solanum melongena L. var. _insanum_ (L.) Prain

Prickly herb; leaveas ovate, pubescent. Flower violet, in racemes. Berries globose, yellow.

Multi-radiate stellate trichomes were observed on both surfaces.

Solanum aculeatissimum Jacq.

An undershrub; prickles long; leaves lobed; small white flowers; berries orange, globose.

Unicellular finger trichomes on both surfaces with gland tipped finger trichomes and glands without necks also in the lower epidermis.

Solanum wendlandii Hook.f.

A climbing shrub; prickles scattered on leaves. Leaves variable; cymes 20-25 cm across. Corolla rotate, lilac coloured.

Trichomes were absent on both upper and lower surfaces.

Discussion

A systematic account of hair morphology in *Solanum* has been given earlier by Seithe (1979). However, the varying degrees of diversity in the trichome characteristics observed in the present study could be correlated with the evolutionary trends of the genus from the morphological stand point of view.

Maximum diversity in trichome types was observed in *S. torvum, S. nigrum, S. melongena, S. pubescence, S. indicum* and *S. macranthum*. On the contrary *S. wendlandii* is glabrous (Table 1). Cowan (1950) considered trichome diversity an index of the evolutionary advancement achieved by the species. The upper epidermis was devoid of trichomes in two species (*S. trilobatum* and *S. spirale). S. melongena* var. *insanum* exhibited only multiradiate stellate hairs.

S. nigrum with herbaceous habit and very small flowers was observed to be the most advanced with bicellular and multicellular finger hairs, multicellular glands, medium sized multi-radiate stellate hairs and unique multicellular finger hairs with constricted cells and very narrow tapering apical cells. Similarly *S. macranthum* characterised by advanced morphological traits such as tree habit and large flowers exhibited abundant diversity in the trichome characters – unicellular and bicellular finger trichomes, milticellular glands, tri-radiate, tetra-radiate and multi-radiate stellate hairs of medium and very large sizes, some of which are stalked.

The genus *Solanum* presents a wide range of variation with many of the members exhibiting up to 5-6 different types of trichomes, while a few are glabrous. Such variations in the trichome features could be correlated with the evolutionary trends from the morphological stand point of view also. The study highlights the significance of trichomes as stable tools in the evaluation of current taxonomic classification.

Table 1. Trichome classes in eleven species of the genus *Solanum*

Species name	Upper epidermis	Lower epidermis
Solanum nigrum	Bi-, multi-cellular finger trichomes Multi-cellular glands Unique multi-cellular finger with constrictions and inflations; end cell tapering.	Bi-, multi-cellular finger trichome Multi-cellular glands Multi-radiate stellate hairs Unique multi-cellular finger with constrictions and inflations; end cell tapering.
S. spirale	No trichomes	Uni-, and bi-cellular finger trichomes Multi-cellular glands.
S. torvum	Uni-, and bi-cellular finger trichomes Multi-cellular gland & short necked trichomes Bi-,tri-,tetra-and multi-radiate stellate hairs (medium, large)	Tri-, and multi-radiate stellate hairs (very large) Multi-cellular glands
S. indicum	Unicellular finger hairs Multi-cellular glands Stalked multi-radiate stellate hairs.	Unicellular and bicellular finger hairs Multi-radiate stellate hairs (large, medium and small) Stalked multi-radiate stellate hairs.
S. pubescens	Unicellular finger hairs Multi-cellular glands Gland tipped finger hairs Multi-radiate stellate hairs	Multi-cellular gland tipped hairs Multi-radiate stellate hairs Gland tipped multi-radiate stellate hairs with multicellular stalks.
S. melongena	Uni-cellular finger trichome Tri-, multi-radiate stellate hairs	Unicellular finger trichome Multi-cellular glands Multi-radiate stalked stellate hairs (medium and small)

S. trilobatum	No trichomes	Multi-radiate stellate hairs (medium)
S. macranthum	Unicellular finger trichomes Multi-radiate stellate hairs	Unicellular finger trichomes Multicellular glands Tri-, tetra-radiate stellate hairs Multi-radiate stellate and stalked stellate hairs
S. melongena L. var. *insanum*	Multi-radiate stellate hairs (thicker & stouter)	Multi-radiate stellate hairs (long & slender)
S. aculeatissimum	Unicellular finger trichomes	Unicellular finger trichomes Multi-cellular glands Glands without necks.
S. wendlandii	No trichomes	No trichomes

References

Ahmed, K.J. 1964. Cuticular studies in Solanaceae. *Canad. J. Bot.* 42: 793-803.

Cowan, J.M. 1950. The *Rhododendron* Leaf. A study of the epidermal appendages. Edinburgh.

Hunziker, A.T. 2001. The genera of Solanaceae. A.R.G. Gartener Verlag. K.G.

Lavania, S. 1990. Trichome morphology in Indian *Solanum*. *J. Ind. Bot. Soc.* 69: 143-148.

Metcalfe, C.R. and L. Chalk 1950. *Anatomy of Dicotyledons.* 1. Clarendon Press. Oxford.

Ogg, A.G., B.S. Rogers and B.E. Schilling 1981. Characterisation of Black Night shade (*Solanum nigrum*) and related species in the U.S. *Weed Sci. Soc. Am. J.*, 29: 27-32.

Seith, A. 1979. Hair types as taxonomic characters in *Solanum*. In Hawkes, J.G., R.N. Lester and A.D. Skelding. ed. *The Biology and Taxonomy of the Solanaceae*. Academic Press, London.

EPIDERMAL AND MORPHOLOGICAL STUDIES IN THE ASTERACEAE

Anitha M., V. Devi Priya and V. Anjana*

Department of Botany, S.N. College, Kollam
* Department of Botany, University of Kerala, Kariavattom.

Abstract

Morphological and epidermal features of fifteen members representing five tribes of the family Asteraceae (Compositae) were studied.

The leaves varied in size, colour, shape, margin and texture, with ***Emilia*** being the most divergent owing to its variable leaf shape, obtuse apex, winged base, yellowish green colouration and glabrous surface. The small florets organized into homogamous or heterogamous capitula, different size and shape classes of leaves, complex patterns of venation and annual or perennial herbaceous habit highlight the advanced condition of the family.

The size of the stomatal complex, type of stomata and the number of subsidiary cells varied considerably. Irregular epidermal cells with undulating walls, abnormal stomata with single or degenerating guard cells and contiguous stomata were observed. ***Elephantopus*** possessed up to five different types of trichomes.

The Hierarchical Cluster Analysis reveals that ***Helianthus*** is very distinct from all the other members, followed by ***Elephantopus*** and ***Emilia***, which form separate groups. Hence it is suggested that ***Helianthus*** may be separated from the tribe Heliantheae and included under a new tribe recent observation of knotted 1- like genes in the active stage in the leaves, stems, and inflorescences of ***Helianthus*** by Tioni *et al.* (2003) based on Northern blot experiments. The placement of ***Emilia*** in a separate tribe Senecioneae is justified in the present analysis. Separation of the sub-tribe ***Elephantopus*** from Vernoniiae based on clustered heads and reduced or modified pappus (Robinson, 1999) gains support in this study.

Introduction

Asteraceae *Dumortier* (Compositae *Giseke*) comprising about 1/10th of the total number of flowering plants includes about 30,000 species grouped under 1500 genera (Herman, *et al.*, 2000). The family is represented in India by 900 species under 167 genera (Hajra *et al.*, 1995).

Earlier taxonomic classifications of the family (Bentham and Hooker, 1873; Clarke, 1876) have been based largely on the exomorphological data. Epidermal characters including the stomata and trichomes have not so far been studied in relation to the existing taxonomic classification of the family. The present study attempts to gather preliminary information regarding the inter relationship between the different species belonging to five tribes of the family.

Materials and Methods

Ten quantitative and ten qualitative morphological characters were collected from 15 plants belonging to five tribes viz. Vernonieae, Eupatorieae, Heliantheae, Anthemideae and Senecioneae of Asteraceae from Kollam. The qualitiative characters of leaf were classified following Hickey (1973). Ten epidermal characters were studied from peelings taken following the methodology standardized by Ahmed (1964). The terminology proposed by Metcalfe and Chalk (1950) for stomata and Ramayya (1962a; 1962b) for trichomes were adopted in the present study. The data collected were subjected to Hierarchical Cluster Analysis (Johnson and Wichern, 2001). A dendrogram illustrating the inter-relationships between the fifteen plants was constructed.

Results and Discussion

The foliar characters showed much variation (Table 1) in size (16.48 cm x 12.87 cm in *Helianthus* to 5.62 cm x 0.72 cm in *Parthenium*), shape (simple lobed or dissected as in *Cosmos*), margin (serrate or lobed), apex (mostly acute but obtuse in *Emilia*), base (variable), texture (pubescent or glabrous) and venation (simple craspedodromous to the advanced palinactinodromous types). *Emilia* exhibited yellowish-green leaves with winged leaf bases.

TABLE 1- MEAN VALUES OF QUANTITATIVE MORPHOLOGICAL CHARACTER IN ASTERACEAE

Character	Vernonia cinerea	Elephantopus scaber	Ageratum conyzoides	Eupatorium odoratum	Parthenium hysterophorus	Eclipta alba	Wedelia calendulacea	Tithonia diversifolia	Helianthus annuus	Spilanthes acmella	Synedrella nodiflora	Cosmos sulphureus	Tridax procumbens	Chrysanthemum morifolium	Emilia sonchifolia
Length of petiole (cm)	1.7	1.35	0.78	0.8	0	0	0.17	1.64	6.54	1.72	1.52	1.61	1.4	2.5	1.52
Length of lamina (cm)	5.72	8.29	6.13	7.32	5.62	7.44	6.32	11.33	16.48	5.68	5.91	11.75	6.1	5.5	4.70
Breadth of lamina (cm)	3.28	2.70	3.8	3.81	0.72	1.71	1.5	4.65	12.87	3.25	3.91	10.62	3.45	4.05	2.54
Area of lamina (cm^2)	10.63	15.36	13.3	14.33	15.13	14.58	7.26	20.76	42.73	15.56	15.53	25.7	9.53	13.15	7.4
Perimeter of lamina (cm)	13.19	16.35	15.01	16.5	20.41	14.98	10.9	24.84	49.92	15.44	15.59	30.21	13.18	14.27	8.26
Length of peduncle (cm)	0.44	0.17	0.57	1.38	0.52	0.47	2.59	6.76	5.14	3.87	0	7.46	14	3.06	2.84
Number of bracts	28.6	6.6	25.4	18.4	5.6	13.06	12.8	21.53	54.26	8.06	13.6	8	4.8	13.7	7.66
Length of bract (cm)	0.2	0.19	0.35	0.78	0.18	0.65	0.45	1.4	5.88	1.11	0.67	1.37	0.25	0.57	0.8
Length of disc floret (cm)	0	0	0	0	0.16	0.28	0.75	1.6	1.59	0.26	0.28	1.63	0.5	0.58	0
Length of ray floret (cm)	0.34	0.7	0.47	1.1	0.24	0.49	1.47	4.85	6.04	0.46	0.66	3.30	0.72	2.01	0.79
Diameter of capitulum (cm)	0.09	0.25	0.42	0.27	0.59	0.54	3.17	4.84	18.93	0.57	0.39	3.81	1.03	5.28	0.36

TABLE 2 – MEAN VALUES OF QUANTITATIVE EPIDERMAL CHARACTERS IN ASTERACEAE

Character		Vernonia cinerea	Elephantopus scaber	Ageratum conyzoides	Eupatorium odoratum	Parthenium hysterophorus	Eclipta alba	Wedelia calendulacea	Tithonia diversifolia	Helianthus annuus	Spilanthes acmella	Synedrella nodiflora	Cosmos sulphureus	Tridax procumbens	Chrysanthemum morifolium	Emilia sonchifolia
Upper epidermis	Length stomatal complex- (U) (µm)	50.26	39.86	39.6	40.66	70.93	59.6	55.28	60.13	59.73	43.2	42.53	39.73	82.13	60.1	60.4
	Breadth of stomatal complex- (U) (µm)	60.93	70.53	38.6	31.86	42.66	52	63.33	69.73	51.06	42.13	38.93	39.73	100.13	50.2	61.86
	Stomatal frequency-U)	26.73	42.33	53.66	20.06	10.2	38.86	23.26	27.86	43.06	25.73	77.6	63.93	20.6	33.2	13.33
	Stomatal Index-(U)	22.61	23.14	26.13	8.36	12.28	20.55	23.53	22.4	32.82	23.93	35.88	26.68	22.72	27.7	13.61
Lower epidermis	Length of stomatal Complex-(L) (µm)	89.46	29.85	32.13	50.1	60.53	61.06	61.73	50	59.7	45.06	60.93	60.8	82.53	49.4	70.53
	Breadth of stomatal (L) (µm) Complex-(L)	80.66	28.8	18.13	49.6	58.93	51.46	80.66	37.86	60.26	33.46	41.06	71.33	90.66	39.3	50.66
	Stomatal frequency(L)	16.27	88.8	12.33	35.26	56.86	9.93	19.53	75.53	59.93	63.8	15.86	42.86	12	43.9	44.8
	Stomatal Index (L)	19.22	52.9	12.49	27.8	41.59	22.32	23.31	33.93	31.84	32.89	22.11	26.4	12.38	35.2	24.46

The solitary inflorescences in *Helianthus* were very large and strikingly attractive, while in many members the small capitula were aggregated into groups of 2-many. The enlarged capitula in brilliant colours attract pollinators more efficiently, there by ensuring maximum cross pollination. In the present study, although primitive characters like brochidodormous venation and simple large leaves were observed in *Helianthus*, alternate phyllotaxy and large inflorescence are to be considered advanced traits. Recent observation of knotted-1 like developmental genes in the active stage in the leaves, stem and inflorescences of *H. annuus* by Tioni *et al.* (2003) support the evolutionarily advanced position of *Helianthus* in relation to all the other members of the family.

Epidermal characters such as the size of the stomatal complex varied considerably (Table 2). The length of the stomatal complex was found to be more significant in determining the stomatal frequency than the breadth. Plants with higher stomatal frequencies showed higher stomatal indices. Majority of the plants possessed anomocytic stomata on both surfaces. The upper epidermis of *Elephantopus, Synedrella, Wedelia* and *Emilia* possessed anisocytic stomata. Arrested stomatal development characterized by degenerating guard cells, single guard cells and contiguous stomata were observed in some members. Maximum variation with regard to the trichomes was exhibited by *Elephantopus* with 5 different types of trichomes, on its upper epidermis, followed by *Helianthus* with 4 types on the lower epidermis. Although two-armed hairs were reported earlier in *Vernonia* (Solereder, 1908) and *Parthenium* (Hess, 1938), the present study showed two armed hairs in *Vernonia* and *Ageratum*.

The Dendrogram depicting the results obtained in the Hierarchical Cluster Analysis (Fig. 1) reveals that *Helianthus* is very distinct from all the other 14 plants when all the data are considered together. At the four cluster level, there are four groups, *Helianthus* forming a single group, a second group consisting of *Emilia,* a third group including *Elephantopus* and a fourth group comprising all the remaining plants.

From the present study it is suggested that *Helianthus* is much distinct from all the other members of the tribe Heliantheae. Recent observations by Tioni *et al.* (2003) support the distinctly advanced evolutionary position occupied by *Helianthus*. Hence it is suggested that *Helianthus* may be separated from the tribe Heliantheae and included under a new tribe. Further the placement of *Emilia* in a separate tribe Senecioneae is justified in the present study.

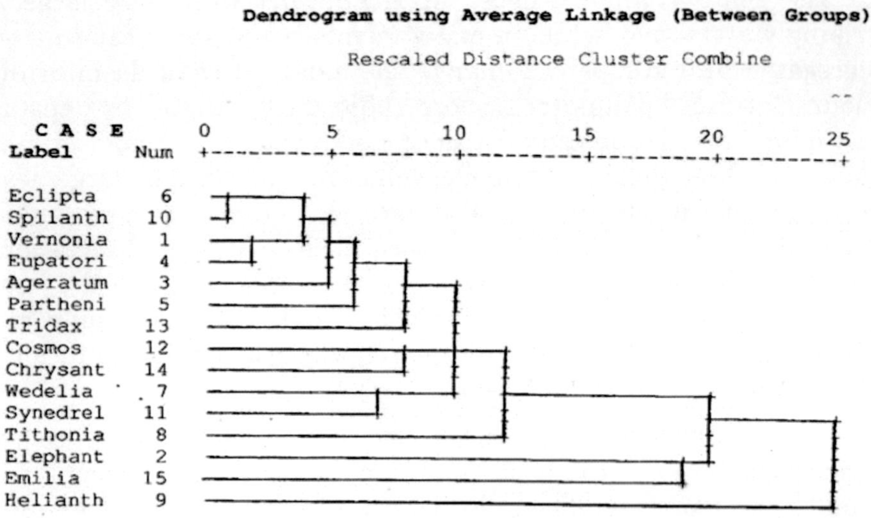

Fig. 1. Response of Chickpea (*Cicer arietinum* L.)

The Hierarchical Cluster Analysis also highlights the gap between *Vernonia* and *Elephantopus* supporting Robinson (1999) who advocated the separation of the subtribe Elephantopodinae from Vernoniiae.

References

Ahmed, K.J. 1964. Cuticular studies in Solanaceae, *Canad. J. Bot.* 42: 793 – 803.

Bentham, G. and J.D. Hooker 1873. Compositae. 163-533. In *Genera Plantarum*, Vol. 2. Reeve and Co., London.

Hajra, P.K., R.R. Rao, D.K. Singh and B.P. Uniyal 1995. Flora of India, Vol. 12 and 13. Botanical survey of India Publications, Calcutta.

Herman, P.P.J., E. Retief, M. Koekemoer and W.G. Welman. 2000. Asteraceae (Compositae). In. Seed Plants of Southern Africa (ed. O.A. Leistner). Strelitzia. 10:101-170. National Botanical Institute, Pretoria.

Hess, R. 1938. Vergleichenda Untersuchungen iiber die Zwillingshaare der composition. *Bot. Jb.* 68: 435 – 496.

Hickey, L.J. 1973. Classification of the architecture of dicotyledonous leaves. *Amer. J. Bot.* 60 (1): 17-33.

Johnson, R.A. and D.W. Wichern. 2001. Applied Multivariate statistical analysis. Prentice Hall, India.

Metcalfe, C.R. and L. Chalk. 1950. *Anatomy of Dicotyledons*, I. Clarendon Press. Oxford.

Ramayya, N. 1962a. Studies on the trichomes of some compositae. II. Phyologeny and classifications. *Bull. Bot. Surv. Ind.* 4:177-188.

Ramayya, N. 1962b. Studies on the trichomes of some compositae. II. Phylogeny and classifications. *Bull. Bot. Surv. Ind.* 4: 189-192.

Solereder, H. 1908. Systematic Anatomy of the Dicotyledons. I. Pp.456-469. Indian reprint (1986). Sushin Printers, New Delhi.

Tioni, M.F., D.H. Gonzalez and R.L. Chan 2003. Knotted 1- like genes are strongly expressed in differentiated cell types of sunflower. *J. Exp. Bot.* 54 : 681-690.

TO FRESH, COMPOSTED, VERMICOMPOSTED *PARTHENIUM HYSTEROPHORUS* AND POULTRY MANURE

B. Vijayakumari and R. Hiranmai Yadav

Department of Botany, Avinashilingam Deemed University, Coimbatore

Abstract

An experiment was conducted at Avinashilingam University Campus, Coimbatore to analyse the impact of fresh, composted and vermicomposted parrthenium and poultry manure and NPK on a legume, *Cicer arietinum*. The above manures were applied in different dosages to the plants. The growth parameters like root length, shoot length, fresh weight, dry weight and vigour index were observed on 30, 60 and 90 Days After Sowing (DAS). The root length was increased due to vermicomposted poultry manure on 30 DAS, vermicomposted parthenium on 60 DAS and composted parthenium on 90 DAS. The decreased root length was in plants treated with fresh poultry manure on 30 DAS, NPK on 60 DAS and control on 90 DAS. The shoot length was more in vermicomposted parthenium application on 30 and 60 DAS, and vermicomposted poultry manure on 90 DAS. The lesser length of shoot was in NPK on 30 DAS, control on 60 DAS and fresh poultry manure on 90 DAS. The fresh and dry weights of plants were maximum in composted poultry manure on 30 and 60 DAS, and composted parthenium on 90 DAS. The vigour was increased due to composted poultry manure on 60 DAS and vermicomposted poultry manure on 30 and 90 DAS. The fresh weight, dry weight and vigour index were observed to be minimum in control on 30, 60 and 90 DAS.

Introduction

The nutrient management in organic farming is concern to within the closed system. Organic materials such as FYM, poultry manure, vermicompost, green manure and crop residue can substitute for inorganic fertilizers to maintain the environmental quality (Chaudhary, 2002).

Organic manures supply the organic matter. The raw organic matter get transformed through the processes of decomposition and humification into humus – the most important organic constituent of the soil which helps to improve the physical, chemical and biological properties of the soil. Soil structure is the key to soil fertility (Kumaraswamy, 2002).

In the face of global ecological concern on the use of chemicals in agricultural production, alternative sources of plant nutrients such as poultry manure are being tested as a source of nutrients to crop plants (Asiegbu and Oikeh, 1995 and Aliyu and Olarewaju, 2000).

Parthenium hysterophorus Linn. is an exotic weed causing severe human health problems as well as agricultural losses. Eco-friendly use of it should be considered to reduce this menace (Renukadevi, 2002).

The present study is an attempt to utilize parthenium and poultry manure composts and vermicomposts as organic fertilizers for chickpea.

Methodology

A pot culture experiment was conducted to study the impact of fresh, composted and vermicomposted parthenium and poultry manure on the biometric characters of chickpea. The experiment was set in Completely Randomised Design (CRD) with three replications. Red soil and sand in 2 : 1 ratio was filled in pots of 7 kg capacity. The chickpea seeds were sown at the rate of 10 seeds / pot.

The treatment details :

T_0	-	Control Red Soil
T_1	-	Fresh parthenium @ 35.0 g / pot
T_2	-	Composted parthenium @ 26.25 g / pot
T_3	-	Composted parthenium @ 35.0 g / pot

T_4	-	Composted parthenium @ 43.75 g / pot
T_5	-	Vermicomposted parthenium @ 26.25 g / pot
T_6	-	Vermicomposted parthenium @ 35.0 g / pot
T_7	-	Vermicomposted parthenium @ 43.75 g / pot
T_8	-	Fresh poultry manure @ 35.0 g / pot
T_9	-	Composted poultry manure @ 26.25 g / pot
T_{10}	-	Composted poultry manure @ 35.0 g / pot
T_{11}	-	Composted poultry manure @ 43.75 g / pot
T_{12}	-	Vermicomposted poultry manure @ 26.25 g / pot
T_{13}	-	Vermiomposted poultry manure @ 35.0 g / pot
T_{14}	-	Vermicomposted poultry manure @ 43.75 g / pot
T_{15}	-	NPK @ 25 : 50 : 0 kg / ha

The biometric characters like root length (cm), shoot length (cm), fresh and dry weights (g) and vigour index were observed on 30, 60 and 90 days after sowing (DAS). The data obtained were statistically analysed (Panse and Sukhatme, 1978).

Results

The biometric characters observed and values obtained are depicted in Table 1.

- **Root length**

The maximum length of root was in T_{13} (20.97 cm) and least in T_8 (9.50 cm) on 30 DAS. On 60 DAS, T_5 (24.23 cm) showed maximum length of root compared to T_{15} (13.78 cm) which had lowest value. On 90 DAS it was increased by T_3 (58.23 cm) and minimum was in control (25.53 cm).

- **Shoot length**

On 30 and 60 DAS, the longest shoot was observed in T_7 (36.37 and 55.00 cm) and shortest in T_{15} on 30 DAS (20.73 cm) and control (28.30 cm) on 60 DAS. On 90 DAS, T_{14} registered the longest shoot length of 58.33 cm and lowest in T_3 (44.83 cm)

Table 1. Impact of fresh, composted and vermicomposted parthenium and poultry manure on biometric characters of *Cicer arietinum* L.

	Root length (cm)			Shoot length (cm)			Fresh weight (g)			Dry weight (g)			Vigour index		
	30 DAS	60 DAS	90 DAS	30 DAS	60 DAS	90 DAS	30 DAS	60 DAS	90 DAS	30 DAS	60 DAS	90 DAS	30 DAS	60 DAS	90 DAS
T_0	11.40	16.17	25.53	20.93	28.30	48.10	1.55	4.37	38.70	0.30	0.84	7.70	1713.33	2361.33	4114.67
T_1	14.70	16.80	33.47	29.07	36.56	49.57	2.56	9.42	60.47	0.47	1.86	12.25	2476.00	3027.33	5086.00
T_2	19.63	20.33	32.07	25.77	37.73	40.73	2.89	9.72	44.95	0.65	2.01	8.99	2726.67	3485.67	4362.00
T_3	18.53	20.70	58.23	25.57	36.40	50.73	3.60	8.93	51.76	0.61	1.77	10.23	2667.67	3228.00	6526.33
T_4	18.97	22.00	47.40	26.80	44.50	52.97	3.14	14.40	76.45	0.62	2.88	15.25	2885.67	4218.33	6349.00
T_5	15.47	24.23	33.27	29.90	50.50	55.40	5.02	14.69	46.92	0.80	4.15	9.25	2725.67	4483.33	5307.00
T_6	20.33	22.70	31.13	28.37	39.33	52.87	5.57	9.52	52.86	1.03	1.89	9.87	3082.33	3927.00	5306.67
T_7	13.83	19.07	46.27	36.37	55.00	53.33	3.97	29.80	52.60	0.85	6.01	9.81	3334.67	4707.67	6640.33
T_8	9.50	14.70	36.20	29.40	42.43	44.83	3.61	28.03	49.05	0.73	5.41	9.74	2588.67	3810.00	5394.00
T_9	19.53	16.50	44.73	25.06	47.30	53.00	3.33	23.32	49.99	0.67	4.46	10.16	2641.00	4479.00	6859.67
T_{10}	12.30	22.17	40.70	31.27	53.30	48.53	3.65	31.57	41.37	0.72	6.49	9.43	3047.00	5273.67	6253.33
T_{11}	14.16	21.33	38.00	32.17	54.10	53.73	6.19	25.48	56.96	1.14	4.71	10.69	3254.00	5277.00	6418.00
T_{12}	18.63	21.27	43.50	28.00	46.17	47.67	4.50	19.42	61.31	0.90	3.34	11.43	3556.67	4925.00	6988.67
T_{13}	20.97	19.83	35.53	28.80	48.17	48.73	3.95	19.31	54.91	0.75	4.11	10.17	3646.67	4992.00	6160.67
T_{14}	14.27	14.80	48.13	28.06	46.20	58.33	4.57	16.31	67.68	0.91	2.14	14.66	2948.67	4272.00	7468.00
T_{15}	9.57	13.73	36.90	20.73	43.87	44.10	2.29	14.17	43.64	0.45	2.86	7.82	1818.00	3456.00	4860.00
SEd	0.88	0.58	3.16	1.58	0.83	1.59	0.32	0.33	2.90	0.10	0.34	1.14	370.12	592.69	771.44
CD (5%)	1.78**	1.17**	6.44**	3.22**	1.69**	3.24**	0.66**	0.68**	5.91**	0.21**	0.69**	2.32**	753.93**	1206.08**	1571.42**

- **Fresh and dry weights**

The fresh and dry weights were maximum in T_{11} (6.19 and 1.14 g) on 30 DAS, in T_{10} (31.57 and 6.49 g) on 60 DAS and in T_4 (76.45 and 15.25 g) on 90 DAS. Control (T_0) plants had least fresh and dry weights of 1.55 and 0.30 g on 30 DAS, 4.37 and 0.84 g on 60 DAS and 38.70 and 7.70 g on 90 DAS.

- **Vigour index**

The increased vigour index was in T_{13} (3646.67) on 30 DAS, T_{11} (5277.00) on 60 DAS and T_{14} (7468.00) on 90 DAS. The vigour index was least in control on 30 (1713.33), 60 (2361.33) and 90 (4114.67) DAS.

Discussion

The root and shoot lengths were improved by vermicomposted parthenium and poultry manure. The fresh and dry weights were maximum in composted parthenium and poultry manure. The vigour index was more in composted and vermicomposted poultry manure.

Similar observations were made by Kavitha Garg and Nagendra Bhardwaj (2000) in wheat plants, where vermicomposted parthenium increased root and stem length of wheat. A possible explanation for the beneficial effect of vermicompost may be the accumulation of mobile substances in earthworm casts. Rajavel (2002) has reported increased dry matter production in cowpea due to composted poultry manure application. This may be due to better nutrition provided by the manure. Neetha Sharma and Mira Madan (1988) observed best growth of maize and wheat with application of poultry waste with and without earthworms.

References

Aliyu, L. and J.D. Olarewaju 2000. Comparative effectiveness of combined application of organic and mineral fertilizers on pepper. *Nigerian J. Hortic. Sci.* 4 : 14-23.

Asiegbu, J.E. and S. Oikeh 1995. Evaluation of chemical composition of manures from different organic wastes and their potential for supply of nutrients to tomato in a tropical ultisol. *Biol. Agric. and Hortic.* 12: 47-60.

Chaudhary, D.R. 2002. Organic farming : An overview. *Farmer's Forum* 2(4): 7-9.

Kavitha Garg and Nagendra Bhardwaj 2000. Effect of vermicompost of parthenium on two cultivars of wheat. *Indian J. Eco.* 27(2): 177-180.

Kumaraswamy 2002. Hi-tech. Agriculture. *Kisan World* 29(3): 41-42.

Neeta Sharma and Mira Madan 1988. Effects of various organic wastes alone and with earthworms on the total dry matter yield of wheat and maize. *Biological Wastes* 25 : 33-40.

Panse, V.G. and P.V. Sukhatme 1978. Statistical methods for agricultural workers, ICAR, New Delhi P. 327.

Rajavel, M. 2002. Integrated nutrient management for sustaining crop productivity and soil fertility under maize based cropping system. Ph.D. Thesis (Agronomy), Tamil Nadu Agricultural University, Coimbatore.

Renukadevi, A. 2002. Protection measures to manage parthenium. *Kisan World* 29 (4) : 55.

STOMATAL VARIATIONS IN CULTIVARS OF *HIBISCUS ROSA-SINENSIS* L.

Devi Priya V., C.A. Ninan* and N.M. Kothari

Department of Botany, S.N.College, Kollam
*Department of Botany,University of Kerala, Kariavattom

Abstract

The stomatal characteristics of sixty cutivars of *H. rosa-sinensis* L. were studied. Epidermal cells were irregular being highly, moderately or less sinuous. The hypostomatic leaves with more undulations in the lower epidermis reflect the xeric adaptation to limit transpiration to the lower leaf surface. The cultivar C-29 possessed amphistomatic leaves.

The stomata were mostly anisocytic (in 73.33% of the cultivars). Stomatal diversity (anisocytic and anomocytic) observed in 16.67% of the cultivars highlights the ongoing process of evolution within the species-complex.

The stomatal complexes showed wide variation in size (length and breadth) ranging from 41.43 μm x 42.9 μm in C-1 to 101.57 μm x 85.43 μm in C-29. Although the stomatal frequency was as high as 34.11 in the cultivar C-1 with the smallest stomata, the remaining cultivars possessed lower stomatal frequencies (<18.5). The stomatal indices ranged from 7.80 in C-8 to 24.21 in C-1.

The analysis of variance showed highly significant variance at 1% level for all the 4 quantitative stomatal characteristics suggesting notable variation among the sixty cultivars for these characters.

Single guard cells and degenerating guard cells were observed in a few cultivars, reflecting stomatal ontogenic aberrations due to various factors.

The *H. rosa-sinensis* species-complex thus exhibits a wide range of variation with regard to its quantitative as well as qualitative stomatal characteristics. The consistency in the pattern of variations of such characters highlights the significance of epidermal morphology as a stable supplementary tool in tackling taxonomic intricacies especially at the lower taxonomic levels.

Introduction

Hibiscus rosa-sinensis L. is one of the most favoured species of the pantropical family Malvaceae with large showy flowers exhibiting much variation in form and shade. This has led to a multiplicity of varieties constituting the *H. rosa-sinensis* species-complex.

The epidermal features such as the epidermal cells, stomata and trichomes are known to yield supplementary clues for solving taxonomic problems (Stace, 1984). Most of the previous studies on epidermal morphology of the family have been restricted to plants at higher taxonomic levels (Inamdar and Chohan, 1969; Rao and Ramayya, 1981). Studies on the epidermal morphology of *H. rosa-sinensis* at the varietal level are lacking.

Materials and Methods

Epidermal studies were carried out on 60 cultivars of *Hibiscus rosa-sinensis* following the methodology standardized by Ahmed (1964). The epidermal characters studied include the distribution of stomata, shape of epidermal cell, nature of epidermal cell wall, type of stomata, number of subsidiary cells, length of stomatal complex, breadth of stomatal complex, stomatal frequency, stomatal index and stomatal abnormalities. The stomata were classified following the terminology proposed by Metcalfe and Chalk (1950). Photomicrographs were taken using an Olympus BH-2 research microscope. The quantitative characters of the lower epidermis were analysed statistically to study the extent of variability among the cultivars of the species, using the analysis of variance (ANOVA).

Results and Discussion

Majority of the cultivars were hypostomatic (Table 3). Cultivar No. C-29 was amphistomatic.

Table 1. Mean values of the quantitative characters of the lower epidermis in cultivars of *Hibiscus rosa-sinensis* L

Cultivar No.	Length of Stomatal Complex (μm)	Breadth of Stomatal Complex (μm)	Stomatal Frequency	Stomatal Index
C-1	41.43	42.9	34.11	24.21
C-2	60.13	59.03	15.55	15.99
C-3	62.7	60.13	18.33	12.81
C-4	69.3	63.43	16.56	18.38
C-5	52.43	52.07	10.56	16.04
C-6	49.5	49.5	16.89	11.05
C-7	60.17	58.67	16.44	16.66
C-8	67.10	64.53	8.56	7.80
C-9	83.97	80.22	8.78	9.77
C-10	66	66	17.33	18.72
C-11	56.1	57.93	15	15.48
C-12	64.17	63.80	10.22	14.61
C-13	55	60.13	17.11	14.66
C-14	80.3	78.83	5.33	12.08
C-15	78.10	77.73	10.67	13.83
C-16	74.07	72.97	9.56	10.06
C-17	49.5	49.5	16	16.10
C-18	65.3	61.97	16.11	19.53
C-19	90.93	76.27	9	13.67
C-20	77.73	76.22	6.89	10.69
C-21	69.37	66	7.67	10.28
C-22	68.57	72.60	12.56	13.32
C-23	69.3	63.07	13.56	18.26
C-24	73.7	70.03	10.11	11.52
C-25	78.1	72.6	11.78	12.68
C-26	73.7	66	6.11	8.33
C-27	81.03	77	8	13.60
C-28	73.33	67.47	10.11	12.14
C-29	101.57	85.43	6.78	10.96
C-30	86.17	68.57	10.67	21.07
C-31	66	66	14.67	19.26
C-32	74.77	63.07	8.89	11.92
C-33	64.9	56.1	17.44	13.91
C-34	71.87	63.07	9.67	10.08
C-35	82.87	74.07	11.33	16.64
C-36	73.7	66	16.22	14.48
C-37	66	56.1	19.67	18.76
C-38	90.57	72.97	5.78	9.79
C-39	67.10	66	10	11.74
C-40	49.5	49.5	16.78	17.95

Table 2. Analysis of Variance of Table 1

C-41	53.9	53.9	15.89	14.29
C-42	82.5	77	8.67	12.89
C-43	70.4	70.4	8	10.98
C-44	70.03	66.73	14.22	19.96
C-45	79.57	73.7	9.22	16.99
C-46	69.67	66	8.11	14.96
C-47	70.03	69.3	10.11	13.07
C-48	69.67	66	10.22	12.66
C-49	92.03	69.67	6.67	7.26
C-50	82.5	66.73	9.67	12.44
C-51	68.57	66	9.33	9.77
C-52	72.6	71.87	10.22	13.53
C-53	78.10	78.47	7.22	15.33
C-54	60.13	53.17	14	17.29
C-55	92.03	77.37	8.89	14.61
C-56	66	66	8.89	11.93
C-57	55.73	55	17.56	14.20
C-58	66	66	10.22	14.29
C-59	69.3	66	9.78	12.95
C-60	79.57	74.07	10.78	14.88

Source	d.f	Mean Sum of Squares			
		Length of stomatal complex	Breadth of stomatal complex	Stomatal Frequency	Stomatal Index
Treatments	59	1272.56**	692.58**	202.46**	105.27**
Error	539	25.39	22.46	1.44	2.82

*significant at 5% level ** significant at 1% level

(i) Epidermal cells

The size, shape, orientation and anticlinal wall undulations of the epidermal cells have been considered as stable taxonomic criteria in the delimitation of taxa (Barthlott and Voit, 1979). All the sixty cultivars of *H. rosa sinensis* included in the present study possessed irregular epidermal cells with more undulations on their walls in the lower epidermis. The lower epidermal cells were either highly (in 38.33% of the cultivars) or moderately sinuous (50%), being less sinuous in only seven cultivars (11.67%). The upper epidermal cells on the contrary were nearly straight-walled in 48.33%, less sinuous (36.67%) or moderately sinuous (15%). The straight walled nature of the epidermal cells has been considered as a xeric character, while the mesomorphic taxa exhibit undulate cell walls (Gbile, 1986). The lesser undulations on the upper epidermis might be attributed to the increased water stress experienced by the upper epidermal cells owing to the exposure to the tropical sun.

(ii) Stomata

The stomatal complexes showed much variation in their length and breadth, ranging from 41. 43 μm x 42.9 μm in the cultivar C-1 to 101.57 μm x 85.43 μm in C-29 (Table 1). The cultivars C-6, C-17 and C-40 also possessed smaller stomata (49.5 μm x 49.5 μm) while C-9 (83.97 μm x 80.22 μm) and C-55 (92.3 μm x 77.37 μm) exhibited larger stomatal complexes. The stomatal frequencies varied from 5.33 in C-14 to 34.11 in C-1. The cultivars C-38 and C-26 also showed lower frequencies of stomata (5.78 and 6.11 respectively). A stomatal frequency as high as 34.11 was noticed in only one cultivar C-1 in which the stomata were the smallest. In all the remaining cultivars, the stomatal frequencies were below 18.5. Stomatal index has been utilized as a dependable morphological parameter in assessing stomatal variability (Salisbury, 1927). In the present study, the stomatal indices ranged from 7.80 in the cultivar C-8 to 24.21 in C-1. A few other cultivars viz. C-26, C-9 and C-38 showed lesser stomatal indices (8.33, 9.77 and 9.79 respectively).

The analysis of variance (Table 2) showed significant variance at 1% level for all the quantitative stomatal characteristics, suggesting wide variation among the sixty cultivars with regard to these characters.

Almost all the cultivars except C-29 were hypostomatic. Rao and Ramayya (1981) have correlated hypostomatic condition with xeric habit. The stomata were mostly anisocytic (73.33%) with 3 subsidiary cells - 2 large and one small. In some cultivars (16.67%) both anisocytic and anomocytic stomata (with 3-6 subsidiary cells) were observed in the same leaf. Inamdar and Chohan (1969) described anomocytic, paracytic and anisocytic stomata in the vegetative and floral organs of *H. rosa-sinensis* and pointed out that the ontogeny of the anisocytic type is syndetocheilic or mesogenous and that of the anomocytic is haplocheilic or perigenous. Bessis and Guyot (1979) have held that such phenomena might reveal an intermediate stage in the evolution of the species. The diversity in the stomata of the same organ, in cultivars of *H. rosa-sinensis* may hence be considered, to reflect the ongoing process of evolution with in the species-complex.

Even though Inamdar and Chohan (1969) have reported single guard cells and contiguous stomata in *H. rosa-sinensis*, the present study showed only single guard cells and degenerating stomata in the various cultivars. Single guard cells were observed in eleven cultivars (18.33%), degenerating stomata in four (6.67%) and both in six cultivars (10%). The stomatal ontogenic aberrations have been attributed to a variety of factors like specific nature (Klebs, 1904), gene action (Mc Clintock, 1956) and momentary developmental disturbances (Rao and Ramayya, 1967).

References

Ahmed, K.J. 1964. Cuticular studies in Solanaceae. *Canad. J. Bot.* 42:793-803.

Barthlott, W. and G. Voit 1979. Mikromorphologic der samen schalen and taxonomic der Cactaceae: Ein rasterelekronen mikroskopischee uberblick – *Plant Syst. Evol.* 132 : 205-229.

Bessis, J. and M. Guyot 1979. An attempt to use stomatal characters in systematic phylogenetic studies of the Solanaceae. In J.G. Hawkes. R.N. Lester and A.D. Skelding (eds.). *The Biology and Taxonomy of the Solanaceae.* Acad. Press, London. pp. 321-326.

Gbile, Z.O. 1986. Epidermal studies in the *Solanum nigrum* complex in Nigeria. In W.G. D' Arcy (ed.) *Solanaceae, Biology and Systematics.* pp. 159-168.

Inamdar, J.A. and A.J. Chohan 1969. Epidermal structure and ontogeny of stomata in vegetative and floral organs of *Hibiscus rosa-sinensis* L. *Aust. J. Bot.* 17: 89-95.

Klebs, G. 1904. Uber probleme der entwicklung. *Biol. Zbl.* 24: 257-267, 287-305, 449-465, 481-501, 545-559.

McClintock, B. 1956. Intranuclear systems controlling gene action and mutation; *Brookhaven Symp. Biol.* 8: 58-74.

Metcalfe, C.R. and L. Chalk 1950. *Anatomy of Dicotyledons – I and II.* Clarendon Press, Oxford.

Rao, R.B. and N. Ramayya 1967. Stomatal abnormalities in two dicotyledons. *Curr. Sci.* 36 : 357-358.

Rao, S.S.R and N. Ramayya 1981. Distribution of stomata and its relation to plant habit in the order Malvales. *Ind. J. Bot.* 4(2) : 149-156.

Salisbury, E.J. 1927. On the causes and ecological significance of stomatal frequency with special reference to the woodland flora. *Phil. Trans. Roy. Soc. B* 216 : 1-65.

Stace, C.A. 1984. The taxonomic importance of the leaf surface. In V.H. Heywood and D.M. Moore (eds.). *Current Concepts in Plant Taxonomy.* London. pp. 67-94.

FOOD HABITS OF TIGERS IN PARAMBIKULAM WILDLIFE SANCTUARY, KERALA

Shijo Joseph*, A.P. Thomas*, R. Satheesh*, R. Sugathan and S. Swaminathan*****

*School of Environmental Sciences, Mahatma Gandhi University
**Ecologist, Parambikulam Wildlife Sanctuary
***Research Biologist, Bombay Natural History Society

Abstract

Parambikulam Wildlife Sanctuary, with an area of 275 sq. km. situated in the Palghat district of Kerala, is one of the seven Level – II Tiger Conservation Unit (TCU) in the Indian subcontinent. Food habit of tigers in relation with other large carnivores' leopard (*Panthera pardus*) and wilddog (*Cuon alpinus*) was studied in Parambikulam Wildlife Sanctuary through scat analysis technique during the period of November 2002 – April, 2003. A total of 103 scats were analysed for tiger, 188 for leopard and 86 for wilddog. The percentage occurrences of different prey species in their diet were founded out. The result showed that large carnivores were mainly depends on deer family for the bulk of their diet. The diversity index and niche breadth values of tiger's diet were smaller than leopard and slightly higher than wilddog. The niche overlap study showed that tiger and wilddog are in good competition in terms of their prey selection.

Introduction

The tiger (*Panthera tigris tigris*), once widely distributed across the riverine grassland and forest of Asia has become highly endangered due to the habitat loss and fragmentation and is now localized only in small isolated forest areas. The conservation of tiger in the constantly and speedy changing, human altered landscape requires a clear understanding of the prey predator dynamism. Food habits of tigers are central to its ecological niche, play an important role in explaining its behavior and population density and may also have an important implication in the life histories of its prey. The study of diet can help to understand the role of a species in the energy flow and nutrient cycle of an ecosystem. It also sets foundation for the understanding of behavior, population dynamics, habitat use and social organization of a species. Although tiger has a major role in stabilizing the ecosystems, there is little information available on the exact distribution, community structure and interaction between herbivores and tiger in most protected areas in Kerala. Therefore the present study was undertaken to analyse the diet of tiger in comparison with leopard (*Panthera pardus*) and wilddog (*Cuon alpinus*) in Parambikulam Wildlife Sanctuary, for assessing the diversity of prey species.

Materials and Methods

Study area — The area selected for the study is Parambikulam Wildlife Sanctuary which is situated in Palghat district, Kerala State, India (between 76°35′ and 76°50′ E and latitudes 10°20′ and 10°26′ N) (Fig. 1) covering an area of 275Km^2. It is part of the contiguous larger area of the forest comprising Anamalai, Nelliampathi, Sholayar, High Ranges and Palani Hills. The average rainfall is 2300 mm and temperature is 30°C. The vegetation includes evergreen, semi – evergreen, moist deciduous, dry deciduous, wet temperate forests (shola) and man made teak plantations. The diverse habitats and strategic locations of Parambikulam make it one of the faunistically richest areas in Kerala. Easa & Balakrishnan (1990) have described the mammals of Parambikulam WLS. According to Wikramanayake *et al.* (1998) Parambikulam is one of the seven Level-II Tiger Conservation Unit (TCU) in South East Asia. The biotic pressure from human habitation is high in the sanctuary.

Methods – The diet of tiger, leopard and wilddog were studied with the help of scat analysis technique. The scats were collected from nine selected paths based on the level of human disturbance (Fig. 1). The classification of human disturbance into high, medium and low (three tracks each) was based on the experience of the first and fourth authors. The scats were sealed in plastic bags and numbered. Excess scats were removed from the path to prevent repeated sampling at next sampling period.

The collected scats were categorised based on size, shape, bone content as well as associated sign like scraping and scenting marks. The diameter of the tiger and leopard scats was measured around the thickest portion with the help of vernier calipers. All the scats were soaked in water, washed thoroughly to separate prey remains like bones, hooves, hairs etc. The hairs are then dried, washed in alcohol for ten minutes and then in xylene. The hairs are then mounted on a slide in xylene and examined under a microscope for characteristic medullary patterns. The prey species is identified from hairs found in scats using the key of Koppikar and Sabnis (1976), Mukherjee (1994) and Easa (1995). The diversity index, niche breadtg and niche overlap were calculated using the following formulas.

Diversity index

The food species diversity in the diet of the three large carnivores is measured using Shannon-weaver (1949) H'- diversity index.

$$H' = - \Sigma pi \ln pi$$

Where,

Pi = proportion of 'i' th item in the total resource used.

Niche breadth

Niche breadth values were based on Levins (1968) measure of niche breadth using the following formula.

$$B = \frac{1}{\Sigma pi^2}$$

Where,

Pi = Proportion of 'i'th item in the total resource used.

This measure was standardized to a scale of 0-1 by using the following formula.

$$B = \frac{(B-1)}{(n-1)}$$

Where,

n = Number of possible resource states.

Niche overlap

The niche overlaps value in between tiger, leopard and wild dog were calculated as described by Schoener (1970) using the following formula,

$$O_{xy} = 1 - \frac{1}{2} \sum_{i=1}^{i=n} / P_{ij} - P_{ik} /$$

Where,

Oxy = niche overlap between species x and y

Pij = proportion of use of i th resources by species j

Pik = proportion of use of i th resource by species k.

n = number of resource states used.

Results and Discussion

Scats were sampled systematically at 15 days intervals along the sampling transect from November 2002 to April 2003. A total of 377 scats (tiger-103, leopard – 188 and wild dog- 86) were collected. Occurrence of various prey species in the scats of tiger, leopard and wilddog were noted (Figs. 2-4). A total of 103 scats were analyzed for tiger among which chital, sambar and gaur constituted about 80% of their diet. A total of 188 scats were analysed for leopard and the result shows that leopard diet is more diverse than tiger and are adaptable to different sizes of prey species. Their preferred food species are chital, sambar and langur. Of the total of 86 scats were analysed among which chital and sambar constituted about 70% of the total prey species consumed.

Niche overlap analysis (Table 1) showed that a strong overlap exist in all the three large carnivores, because all are depend on two or three prey species for their major diet. Tiger and wilddog showed higher overlap (0.84) followed by tiger and leopard (0.75) and leopard and wilddog (0.71). All niche overlap values were higher than (0.5) indicating that all the large carnivores species are in competition with each other and our results supports the study of Karanth and Sunquist (1995) in Nagarhole National Park and Arivazhagan (1997) in Sigur and Thalamala Reserve Forest. The niche breadth hierarchy

(Table 2) was in the order of leopard (0.57) > tiger (0.5) > wilddog (0.48). leopard showed high niche breadth value because of their wider access to different sizes of prey species.

Fig.2 - Percentage Occurance of different prey species in the scats of tiger in Parambikulam Wildlife Sanctuary

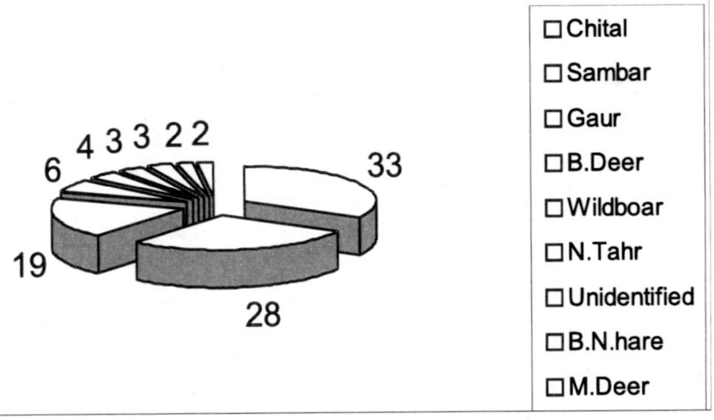

Table 1. Food niche overlap between three predatory species (tiger, leopard and wild dog) in Parambikulam WLS during the study period Nov. 02-Apr. 03 (Based on scat analysis)

S. No	Predatory species	Niche overlap values
1	Tiger X Leopard	0.75
2	Tiger X Wild dog	0.84
3	Leopard X Wild dog	0.71

Table 2. Food niche-breadth values for tiger, leopard and wild dog in the study area during the study period Nov. 02 – Apr 03. (Based on scat analysis)

S. No	Predator Species	Niche breadth values (B)	Standard niche breadth values (B)
1	Tiger	5.007	0.5
2	Leopard	6.690	0.57
3	Wild dog	4.381	0.48

The Shannon - Weaver diversity index of the diet of large carnivores (Table 3) leopard diet (2.11) is more diverse than tiger (1.82) and wilddog (1.68) and it supports Sankhala (1977) that leopards are versatile animals and their diet include a variety of smaller animals and avoid competition with tiger.

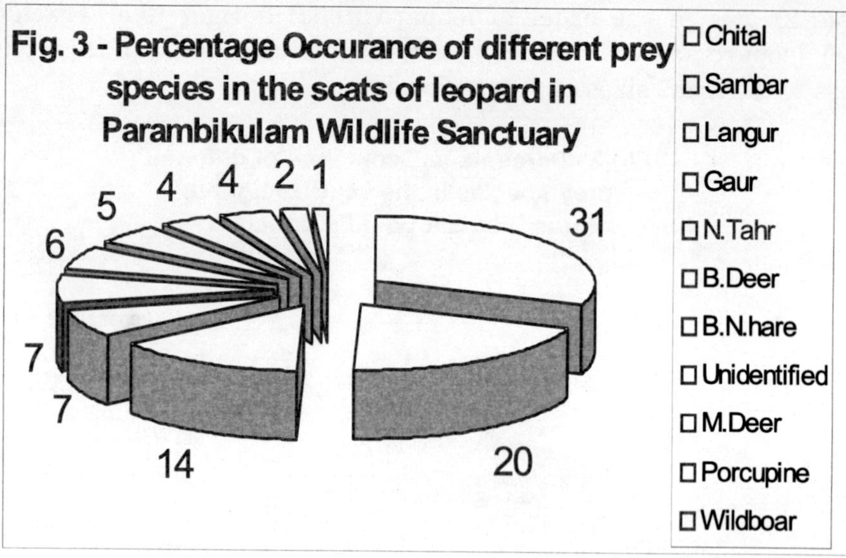

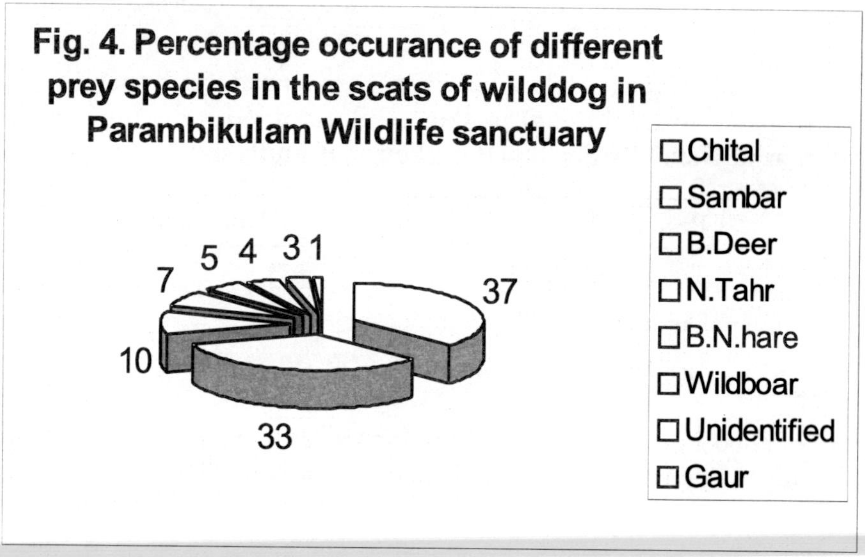

Table 3. Food species diversity index (Shannon & Weaver, 1949) in 3 predatory species during the study period (Nov. 02 – Apr. 03)

S. No	Predator species	Shannon's diversity index. H'
1	Tiger	1.82
2	Leopard	2.11
3	Wild dog	1.68

Conclusion

The foraging ecology of large carnivores in a human dominated landscape showed that all of them preferring deer family for the bulk of their diet. So for the conservation of large carnivores it should be necessary to protect the deer family and its habitat. The diversity index and niche breadth values of tiger's diet were smaller than leopard and slightly higher than wilddog. The niche overlap study showed that tiger and wilddog are in good competition in terms of their prey selection.

Acknowledgement

We are grateful to the Kerala Forest Department for permission and support to do the fieldwork, to the Director, School of Environmental Sciences, Mahatma Gandhi University for all the support to the study and to Dr. R. Satheesh, Dr. R. Sugathan and Mr. Swaminathan for their useful guidelines.

References

Arivazhagan, C. 1997. A study on the food habits of leopard (panthera pardus) in the Sigur Reserve forest of Tamilnadu with special reference to human leopard conflict. M. Phil. dessertation submitted to Bharathithasan University.

Easa, P.S. 1995. Prey predator studies in Eravikulam National Park. KFRI Research Report No. 105, Peechi, Kerala.

Easa, P.S. and M. Balakrishnan 1990. Population ecology and management problem of large mammals in the Parambikulam Wildlife Sanctuary, Kerala. In: Daniel, J.C. and J.S. Serrao (Eds.). Conservation in Developing countries: Problems and Prospects. Proceedings of the centenary seminar of the Bombay Natural History Society, Bombay.

Edgonkar, A.J. and R. Chellam 1998. A Preliminary study on the 3 ecology of the Leopard *Panthera pardus fusca* in the Sanjay Gandhi N.P, Maharashtra. Final Report Wildlife Institute of India.

Garla, R.C., E.Z.F. Setz, and N. Gobbi 2001. Jaguar (*Panthera onca*) food habits in Atlantic rainforest of South-eastern Brazil. *Biotropica,* 33(4): 691-696

Karanth, K. U. and M.E. Sunquist 1995. Prey selection by tiger, leopard and dhole in tropical rainforest, *J. Animal Ecol.,* 64(4): 439-450.

Koppikar, B.R. and J. H. Sabnis 1976. Identification of hairs of some Indian mammals. *J. Bombay Nat. Hist. Soc.* 73: 5-20.

Levins, R. 1968. Evolution in changing environment: some theoretical explorations, Princeton University Press, Princeton, NJ.

Mills. M.G.L. 1992. A comparison of methods used to study food habits of large African carnivores. In Wildlife 2001: populations: 1112 - 1124. Mc Coullough, D. R. and Barret, R.H. (Eds.). London and New York: Elsevier Applied Science.

Mukherjee, S., Goyal, S.P. and Chellam, R. 1994. Standardisation of scat analysis techniques for leopard (*Panthera pardus*) in Gir National Pak, Western India, *Mammalia*, 58(1):139-143.

Sankala, K. 1977. Tiger! The story of the Indian tiger, Simson and Schuster, New York.

Schoener, R.W. 1970. Non-synchronous spatial overlap of lizards in patchy habitats. *Ecology,* 51: 408-418.

Shannon, C.E. and W. Weaver 1949. The mathematical theory of communication, University Illinoise Press, Urbana, IL.

Vernes, K., A. Dennis, and J. Winter 2001. Mammalian diet and broad hunting strategy of the Dingo (*Canis familiaris dingo*) in the wet tropical rainforest of Northeastern Australia. *Biotropica,* 33(2): 339-345.

Wikramanayake, E.D., E. Dinerstein, J.G. Robinson, U. Karanth, A. Rabinowitz, D. Olson, T. Mathew, P. Hedae, M. Konner, G. Hemley and D. Bolze 1998. An Ecology based method for defining priorities for large mammal conservation: The Tiger as case study. *Conservation Biology,* 12 (4): 865 - 878.

MANGROVE FOREST OF KERALA COAST: ITS STATUS, DISTRIBUTION AND MANAGEMENT

P. Christudas George

Department of Botany, Mar Thoma College, Thiruvalla

Abstract

Mangrove forest is a natural asset of any country. But due to unscientific approach to achieve development man is threatening the existence of it. This paper considers the status, distribution and management of mangrove forests in Kerala. By focusing on the various management aspects emphasis is given to aspects that can be used save the diminishing mangrove forests of Kerala.

Introduction

Mangroves are basically perennial plants that grow in coastal areas. They are found in the intertidal zones of Sheltered shores, estuaries, creeks, backwaters, lagoons, marshes, and mud flats. Mangrove "is a tree, shrub, palm or round fern, generally occurring one and a half meter in height and which normally grows above mean sea land in the intertidal zone of marine coastal ecosystem or estuarine margins" (Duke, 2002). The Mangrove vegetation basically consists of evergreen trees and shrubs, belonging to different families with similar physiognomy. They are also similar in their physiological and structural adaptations. Most of them have pneumatophores and stilt roots. They show a marked tendency to vivipary for successful regeneration.

According to the Ministry of Environment and Forest, Government of India (Anonymous 1987), the total area of mangrove left in India is about 6740 Km^2 and its forms only about 7% of that of the world. Kerala has 590 Km long narrow coastal line, constituted by a long stretch of back waters constitutes just about 1- 2% of the total area of India (Basha, 1991). In Kerala a series of lagoons run parallel to the sea separated from back waters by a strip of land. The enrichment of 30 estuaries and back waters with regular supply of fresh water from 41 perennial rivers create peculiar ecological environment leading to the development of a unique mangrove vegetation on the fringes of the back waters estuaries and creeks. This unique flora with their specialised ecological characters affords suitable atmosphere for a set of peculiar Fauna-both living in their own highly specialised ecosystem. The replacement of natural vegetation for need based agricultural and industrial development led to the annihilation of the vast mangrove vegetation in Kerala (Mohanan, 1984).

Distribution of mangroves

A district wise distribution showing approximate extent of mangroves is given below. Of the total area mentioned, which is about 1671 ha (Basha, 1991), about 1470 ha are with the private land holders and only about 200 ha with government agencies. The distribution is purely discontinuous and in patches of varying extent.

THIRUVANTHAPURAM

- Estimated Area covered – 23 ha
- Sub-Regions – Veli, Thumba, Edava, Akkulam, Kandinamkulam.
- Principle Species – *Sonneratia, Avicenia, Bruguiera, Ardisia, Acrostichum.*
- Remarks - A specimen of *Manilkara hexandra* typical to littoral forest of east coast are found here.

KOLLAM

- Estimated Area covered – 58 ha
- Sub-Regions–Asramam, Neendakara, Ashtamudi Sanctuary Ayiramthengu.
- Principle Species – *Cerbera, Avicennia, Rhizophora, Bruiguiera & Lumnitzera.*

- Remarks – Kollam is the only place from where *Lumnitzera racemosa* is reported.

KOTTAYAM

- Estimated Area covered – 80 ha Sub-Regions – Vaikom, Vechoon, Kumarakum.
- Principle Species – Sonneratia, Ardisia, Phragmites.

ALLAPUZHA

- Estimated Area covered – 90 ha.
- Locations – Pathiramanal Island, Chertala, Vayalar, Chanthirur.
- Principle Species – *Cerebera, Samadera.*

ERANAKULAM

- Estimated Area covered – 260 ha
- Locations – vypin, Panagad, Puthuvaippu
- Principle species – *Avicennia, Bruguiera, Rhizophora, Acanthus.*
- Remarks – to conserve mangrove forest Mangalavanam concept is accepted here.

THRISSUR

- Estimated Area covered – 21 ha
- Locations – Chetwai, Theripayar and Kodungallur.
- Principle Species – *Avicennia, Bruguiera, Rhizophora* and *Acanthus.*

MALAPPURAM

- Estimated Area covered – 12 ha.
- Locations – Tirur, Kadalundy.
- Principle Species -*Avicennia, Bruguiera.*

KOZHIKODE

- Estimated Area covered – 293 ha.
- Locations – Nadakavu, Mahe, Beypore.
- Principle Species - *Avicennia, Rhizophora, Bruguiera.*

KANNUR

- Estimated Area covered – 755 ha.
- Locations – Thalassery, Dharmadam, Taliparambu, Pazhayangadi.
- Principle Species – *Avicennia, Sonneratia, Kandelia, Acrosticum.*
- Remarks – Richest district in mangrove in the state.

KASARKODE

- Estimated Area covered – 79 ha
- Location – Nileswaram, Bekal, Manjeswar, Shiria, Mogral
- Principle Species – *Rizhophora, Aegiceras, Avicennia.*

Significance of Mangroves

The mangrove eco-system holds and stabilizes the environment from erosion and acts as a buffer zone between the land and sea. It allows the deposition of metal ions. In biological terms mangrove eco-system is primarily a source of organic matter and acts as a sink. The roots of mangroves trap sediments with their prop roots, thus functioning as land builders and also as a barrier to control the erosion of coastal areas.

Mangroves behave as a purification of efflueuts as the plant absorb the excess nitrates and phosphates, there by preventing the contamination of estuaries. The wide ranges of economic, ecological and cultural aspects of mangroves are yet to be understood fully, but a lot of research has given a great amount of insight on it.

- Mangroves serve as shoreline protectors against tidal surges, cyclonic storms and reduce high velocity wind. The root system of mangroves trap debris and help in soil formation.
- Provides habitat and breeding ground for wild-life, especially migratory birds.
- Mangroves also provide appropriate eco-system and refuge for fish and marine invertebrates, molluscs etc. They also act as sanctuaries to protect the genetic diversity of various organisms.
- Mangroves selectively bind toxic contamination.
- Mangrove forests offer basic need like wood, food, medicines, tannins and construction material to the coastal population.

- Mangroves are very important due to the recycling of nutrients and the enrichment of coastal waters thereby supporting the benthic population.

Status of mangroves vegetation

In Kerala mangroves were grouped based on the species composition and growth status (Basha, 1992).

Fully Mangrove Areas

Here mangrove species were dominant and if it is left undisturbed these places would develop into luxuriant mangrove formation – About 1.5 Km^2 area is under this category. Instances of this kind are seen in Mallikkad in Kasarkode district.

Potential Mangrove Location

Potential mangrove areas are lands where mangrove species are in a condition where the species are persisting and regenerating. Such areas are observed in Pazhayangadi, Baliapatam and Puthurypa. It is estimated that nearly 9 Km^2 of area is under this category.

Areas Under succession

Area under primary and secondary succession where pioneer species like *Acanthus ilicifolis* and *Acrosticum aureum, with* occassional growth of species like *Avicennia, Rhizophora* and *Kandelia* are in the process of establishing mangrove vegetation. The total area under succession is comes about 5-7 Km^2 . Areas like Neendakara in Quilon are example.

Threatened Mangrove Location

Due to prevention of free-flow of tidal water from sea, the chemical composition of the water gets changed. Pollution due to effluents from natural and artificial sources also causes degradation to mangroves. Such areas can be seen in Kollam, Kottayam, Eranakulam and Allapuzha district of Kerla state.

Areas under arrested growth

Growth of trees and species is arrested due to the repeated looping of branches for aqua cultural purposes. It remains as a stunted patch. Such patch is found in the fish farm of Ayiramthengu in Kollam.

Mangement of mangroves

Mangrove a threatened eco-system

The sustained exploitation of mangroves for domestic and Industrial need is considered as the major cause that contribute to mangrove forest destruction. Factors that lead to the annihilation of mangrove forest can be grouped into two - Natural and Artificial.

- **Natural Causes** - Changes in topography and configuration in the coast-line that causes shore line erosion.
- **Artificial Causes** - Deforestation and diversion of land for agricultural and aqua culture, industrial pollution and over exploitation of land for establishing settlements etc are some of the main causes. The after effects have been witnessed in places like Vembabad and Kuttanad region.

Another major reason for the destruction of Mangroves is the lack of awareness among the public and stake holders regarding the importance of mangrove.

Mangrove Conservation

Mangroves are pieces of nature's gift worth preserving. Priority should be given to bio-diversity. Conservation and ecosystem restoration in mangrove area management conservation includes some of these steps.

- Afforestations and reforestation of damaged mangrove areas.
- Coastal zone management policy for prevention of encroachment can be discussed. Mangrove vegetation could be brought under the network of wild life protection areas Act in Kerala.
- A large chunck of mangrove in the state is with private owners. It should be procured from these occupants and conserved.
- Non-consumptive tourism generates the revenue and the motto of conservation should be is upheld by working with nature.
- Traditional Industries near mangrove areas should be mechanised and utilization of space is reduced. Activities related with coir retting is a good example.
- Ecological impacts should be studied before taking up development in mangrove areas.
- Joint mangrove management programmes should be conducted with the co-ordination between the locals and Government agencies.

Conclusion

Suggestions are herewith put forward for the management of mangrove eco-system.

- Mechanized small scale industries near mangrove area, so that area covered by these industries can be decreased.
- Establishment of Ports for fishing and trade.
- The shrinkage of mangrove vegetation to small isolated pockets make the protection difficult. So its area of coverage should be increased by afforestation.
- To devise a national wet land law to protect wetlands.
- Establish of the concept of non-consumptive tourism and spread eco-education programme in the area.
- Restore of a Mangrove Ecosystem Information Services by providing easy on-line access.
- Introduce Remote sensing of mangrove areas to have a on-hand view of development of mangrove areas.
- Mangrove Genetic Resources Centers should be established to improve genetic diversity.
- The mangrove conservation programme should not be just donor driven alone. It should be long-term and objective-driven.

Acknowledgement

I am deeply indebted to my teachers, Dr. Mary Chacko and Dr. Hari N. Iyer for all the valuable support and guidance while preparing this review. I wish to thank Dr. K.K.N. Nair, Scientist Department of Botany, KFRI, Peechi, for his help and also Dr. Swarupanandan, Scientist, Department of Soil-Science KFRI, Peechi, for his help in securing references. Also I am indebted to the Librarian of KFRI, Peechi, for helping me to procure the reference materials.

I also wish to thank Professor Thomas Mathew Head of Department of Botany, Mar Thoma College, for his encouragement through out the preparation of this paper.

References

Anonymous, 1987. *Mangroves in India-Status report,* Government of India, United Printing Press, New Delhi.

Anupama, C. 2004. Mangroves- The Miraculous coast guards. Publication on the training workshop on conservation of mangroves eco-system. Kannur, Kerala. pp. 131-132

Basha, S.C. 1991. Distribution of Mangroves in Kerala. *Indian Forester* 117 : 439-448.

Basha, S.C. 1992. Mangroves of Kerala – a fast disappearing asset. *Indian Forester* 118: 175-189.

Chelliah, S. 2002. The mangrove decade and beyond. Activities lessons and challenges in mangrove conservation and management 1990-2001. M.S. Swaminathan Research Foundation, Chennai. pp. 1-40.

Duke, C.N. 2002. Extracts from The Mangrove the decade and beyond. Activities and lessons and challenges in mangrove conservation and management. M.S. Swaminathan Research foundation, Chennai.

Gamble, G.S. 1915-1935. *Flora of Presidency of Madras* 3 vols. Adland and Sons Ltd., London.

Mohannan, C.N. 1997. The mangroves, In: Natural Resources of Kerala WWF Trivandrum, pp. 149-158.

Ramachandran, K.K. and C.N. Mohannan 1990. The mangrove eco-system of Kerala in mapping inventory and sum environmental aspects, final report CESS, Trivandrum 100 Pgs.

Rekha, R. 2004. The mangroves highly vulnerable eco-system, publications from the training workshop on the conservation of mangrove eco-system, Kannur. pp. 129-130.

A STUDY ON THE AVIAN FAUNA OF AYMANAM PANCHAYATH

Rakesh P.S., S. Prasanth Narayanan and A.P. Thomas

School of Environmental Sciences, Mahatma Gandhi University, Kottayam

Abstract

Avian faunal diversity of Aymanam Panchayath, eastern part of Vembanad wetland ecosystem, was studied during the period of Nov-2003 to Jan-2004. Twelve sampling sites were surveyed by direct observation method using 7 X 50 mm binoculars. A total of eighty-two species of birds belonging to 38 families and 11 orders were identified in which family Corvidae showed maximum diversity of 10 species. Order Passeriformes showed maximum family diversity of 13 nos. Sixteen species of migratory birds, including both the local, Himalayan foot hill and intercontinental Eurasian migrants were identified. The presence of Darter, a Near Threatened species, reveled the importance of Vembanad wetland ecosystem as a Ramsar site. It is recommended that further intensive studies are needed to make the complete checklist of the bird species of the Aymanam Panchayath.

Introduction

Birds are feathered, air breathing, warm blooded, oviparous, bipedal vertebrates. These are the most highly specialized craniates coming under the class – Aves and the phylum Chordata. Almost every part of their organization is modified for aerial life. The distribution of birds around the world is phenomenal. Birds are found in almost every geoclimatic zone. There are 8,600 species of birds in the world, in which 1,200 species were found in India, which accounts for about one – seventh of the world birds population. Being a part of Western Ghats - one of the 25 hotspots identified around the world, Kerala's avian fauna is very rich. 483 species of birds are identified from Kerala includes, 16 endemic bird species of Western Ghats (Rahmany, 2002). The present study aims to know the diversity of birds in the eastern parts of Aymanam Panchayath.

Methodology

Eastern part of the Aymanam Panchayath was selected as study area and surveyed for birds. Twelve sampling sites were fixed in different locations all along the study area. Simple direct observation technique was used for the survey. Observations were done with the help of a 7 x 50 mm binocular. Different habitats of birds from the sampling sites are observed thoroughly. Identification was done with the help of standard reference materials (Grimmet *et al.,* 2000, Grewal *et al.,* 2002, Ali, 1978, Neelakantan, 1996)

Result and Discussion

A total of eighty two species of birds belonging to 37 families in 11 orders were identified from the twelve sites (Table 1). In the present study family Corvidae showed maximum diversity of 10 species (10.1%) and families Ardeidae with 9, Accipitridae, Sylvidae and Nectarinidae came in the next positions, with 4 species each. 20 families are represented by only one species each (Table 1). Order Passeriformes showed maximum family diversity of 13 (34.2%) and order Ciconiformes came second, which had 9 families (22.8%) (Table 2). Of the 82 species identified 27 species (34.1%) were wetland dependent species (Jayson, 2002). Major seed distributors of the dry evergreen forest like *Megalaima viridis* (Smallgreen Barbet), *Megalaima haemacephala* (Coppersmith), *Eudynamis scolopacea* (Koel), *Corvus* splendens (House Crow), *C. macrorhynchos* Jungle Crow, *Dendrocitta vagabunda* (Indian Treepie), *O. oriolus* (Golden Oriole), *Pycnonotus cafer* (Redvented Bulbul), *Accridotheres fuscus* (Jungle Myna), *Turdoides striatuus* (Jungle Babbler), *Dicaeum erythrorhynchus* (Tickelle's Floweepecker) (Balasubramanian *et al.,* 2003) were also identified from the study area *Merops leschenaulti* (Chestnut headed Bee-eater) is a forest species is also spotted from the Site Mariathuruthu. It commonly inhabits stream and pool margins of forests in small groups. Among the 82 species identified *Anhinga melanogaster* (Darter) (*Grewal et al.*, 2002) is a globally threatened species. The *Anhinga melanogaster* (Darter) is considered a Near threatened species, with biogeographical population of as low as 10,000. The 1% threshold of this population is 100. Therefore any wetland having more than 100 Darters on a regular basis could be considered as an Important Bird Area (Rahmany, 2002). *Psittacula krameri* (Roseringed Parakeet), *Glaucidium* radiatum (Jungle Owlet), *Haliastur indus* (*Brahminy* Kite), *Circus aeruginosus* (Western Marsh Harrier), *Accipiter* badius (Shikra) are included in the rare birds list in the book "*A Brief Outline of the Biodiversity Conservation Strategy and Action Plan for Kerala*" 2003. Most of the endemic and endangered birds are living in the forests, so in the study period any endemic bird could not been sighted.

Sixteen species of migratory birds were identified in the present study (Table 3), including both the local, Himalayan foot hill and intercontinental Eurasian migrants. *Gelochelidon nilotica, Actitis hypoleucus, Circus aeruginosus, Plegadis falcinellus, Accrocephalus dumetorum, Phylloscops trochiloides* etc. are considered as intercontinental migrants and *Dendronanthus indicus, Phylloscops magnirostris, Hirundo rustica, Hirundo daurica, Sturnus m. malabaricus, Terpsiphone paradisi, Oriolus oriolus, Pitta brachyura, Childonias hybridus etc* are considered as Himalayan foot hill migrants and *Bubulcus ibis, Ciconia episcopus, Threskiornis melanocephalus* etc are local migrants. But species like *Actitis hypoleucus , Circus aeruginosus, Phylloscops trochiloides, Phylloscops magnirostris, Hirundo rustica, Hirundo daurica* etc also breeds in the Himalayan regions (Bates and Lowther, 1991) and *Dendronanthus indicus, Plegadis falcinellus, Phylloscops magnirostris, Hirundo rustica, Hirundo daurica, Sturnus m. malabaricus, Terpsiphone paradisi, Oriolus oriolus, Pitta brachyura, Childonias hybridus* etc alsosome times migrate to out side of India to Eurasia. *Phylloscops trochiloides, Accrocephalus dumetorum* are seen in all the twelve sites and *Gelochelidon nilotica, Actitis hypoleucus, Plegadis falcinellus, Threskiornis melanocephalus, Pitta brachyura,* unidentified *Accrocephalus spp.* and *Phylloscops magnirostris* are seen only one of the different sites. Family Sylvidae showed maximum migratory species of 3.

Nesting activities of five species of birds were identified from four sites, Pulikkuttissery, Prappuzhakkadavu, Kallumada and Mariathuruthu (Table 4). *Megalaima viridis* (Smallgreen Barbet), *Psittacula krameri* (Roseringed Parakeet), *Corvus splendens* (House Crow), *Parus major* (Grey Tit) and *Prinia inornata* (Plain Wren Warbler) are five bird species and all of them comes under the order Passeriformes. *Megalaima viridis* (Smallgreen Barbet) and *Psittacula krameri* (Roseringed Parakeet) were located from the Prappuzhakkadavu in a dead coconut tree, *Prinia inornata* (Plain Wren Warbler) nested in a *Cyperaceae* family grass species near paddy field, *Corvus* splendens (House Crow) nested in the tree *Thespesia populnea and Parus major* (Grey Tit) nested in the exotic tree *Glyricidia sepium.*

Forty-five species of birds were identified from the western part of the Aymanam Panchayath (Rakesh, 2003); in the present study 82 species could be spotted from the eastern part of the Aymanam Panchayath. Increase in species in the eastern part is due to the winter season. In winter season numerous Eurasian and Himalayan migrants have visited the place. Study areas are wetland dominated, wetlands in Kerala come under Central Asian- Indian flyway. During their annual migrations, water birds halt at sites for varying periods to rest and feed, which act as a 'stepping stones' for their survival. The migratory birds use wetlands of Kerala as 'stepping stones' (Jayson, 2002). A number of Scolopcidae and Charadridae family species were counted in the western Aymanam Panchayath (Sreekumar, 2002) in the winter season. But in the present study only one Scolopacidae family species was spotted owing to the flooded conditions in the paddy fields. The paddy fields in the growing and flooded condition and the presence of numerous trees and thickets might be the reasons for the high diversity of birds. It is recommended that further intensive studies are needed to make the complete checklist of the bird species of the Aymanam Panchayath.

Systematic Checklist of the Birds of Eastern Aymanam Panchayath

Order: Anseriformes

Family: Anatidae

1. Dendrocygna javanica Horsfield — Lesserwhistling Teal
2. *Nettapus coromandelianus.* Gmelin — Cotton Teal

Order: Piciformes

Family: Picidae

3. *Dinopium benghalense* Whistler & Kinnear Lesser GoldenbackedWoodpecker

Family: Megalaimidae

4. *Megalaima viridis* Boddaert	Smallgreen Barbet
5. *M. haemacephala* Latham	Coppersmith Barbet

Order: Coraciformes

Family: Alcedinidae

6. *Alcedo atthis* Kleinschmidt	Common Kingfisher

Family: Halcyonidae

7. *Halcyon capensis* Linnaeus	Storkbilled Kingfisher
8. *H. smyrnensis* Boddaert	Whitebreasted Kingfisher

Family : Cerylidae

9. *Ceryle rudis* Whistler & Kinnear	Pied Kingfisher

Family : Meropidae

10. *Merops philippinus* Linnaeus	Bluetailed Bee-eater
11. *M. leschenaulti* Vieillot	Chestnutheaded Bee-eater

Order : Cuculiformes

Family : Cuculidae

12. Eudynamis scolopacea Linnaeus	Koel

Family : Centropopidae

13. *Centropus sinensis* Streseman	Crow Pheasant

Order : Psittaciformes

Family : Psittacidae

14. *Psittacula krameri* Bechstein	Roseringed Parakeet

Order : Apodiformes

Family : Apodidae

15. *Cypsiurus balasiensis* J.E.Gray	Asian Palm Swift
16. *Tachymarptes melba* Koelz	Alpine Swift
17. *Apus affinis* J.E.Gray	House Swift

Order : Strigiformes

Family : Strigidae

18. *Glaucidium radiatum* Blyth — Jungle Owlet

Order : Columbiformes

Family : Columbidae

19. *Columba livia* Strickland — Blue Rock Pigeon

Order : Gruiformes

Family : Rallidae

20. *Amaurornis phoenicurus* Pennant — Whitebreasted Waterhen
21. *P.Porphyrio* Latham — Purple Moorhen

Family : Scolopacidae

Sub Family : Tringinae

22. *Actitis hypoleucus* Linnaeus — Common Sandpiper

Family : Charadridae

23. *Vanellus indicus.* Boddaert — Redwattled Lapwing

Order : Ciconiformes

Family : Jacanidae

24. *Metopidius indicus* Latham — Bronzewinged Jacana

Family : Laridae

Sub Family : Larinae

Sternini

25. *Gelochelidon nilotica* Gmelin — Gullbilled Tern
26. *Childonias hybridus* Stephens — Whiskered Tern

Family : Accipitridae

Sub Family : Accipitrinae

27. *Haliastur indus* Boddaert — Brahminy Kite
28. *Circus aeruginosus* Linnaeus — Western Marsh Harrier
29. Accipiter badius Gmelin — Shikra

Family : Anhingidae

30. Anhinga melanogaster Pennant	Darter

Family : Phalacrocoracidae

31. *Phalacrocorax carbo.* Shaw	Great Cormorant
32. *Phalacrocorax niger* Vieillot	Little Cormorant
33. *P. fuscicollis* Stephens	Indian Shag

Family : Ardeidae

34. *Egretta grazetta* Linnaeus	Little Egret
35. *Casmerodius albus.* Linn	Large Egret
36. *Ardea purpurea* Mayen	Purple Heron
37. *Mesophoyx intermedia* Wagler	Intermediate Egret
38. *Bubulcus ibis* Boddaert	Cattle Egret
39. *Ardeola grayii* Sykes	Indian Pond Heron
40. *Butoridus striatus* Bonaparte	Little Green Heron
41. *Ixobrychus cinnammomeus* Gmelin	Chestnut Bittern
42. *Dupetor flavicollis* Latham	Black Bittern

Family : Ciconidae

43. *Ciconia episcopus* Boddaert	Whitenecked Stork

Family : Threskiornithidae

44. Threskiornis melanocephalus	Latham White Ibis
45. *Plegadis falcinellus* Linnaeus	Glossy Ibi

Family : Podicipedidaea

46. *Tachybaptus ruficollis.*Salvadori	Little Greebe

Order : Passeriformes

Family : Pittidae

47. *Pitta brachyura* Linnaeus	Indian Pitta

Family : Irenidae

48. *Chloropsis cochinchinensis* Blyth	Bluewinged Leafbird

Family : Corvidae

Sub Family : Corvinae

Corvini

49. *Dendrocitta vagabunda* Whistler & Kinnear	Indian Treepie
50. *Corvus splendens* Madarasz	House Crow
51. *C. macrorhynchos* Sykes	Jungle Crow

Artamini

52. *Artamus fuscus* Vieillot	Ashy Shrike Swallow

Oriolini

53. *Oriolus xanthopterus* Linnaeus	Blackheaded Oriole
54. *O. oriolus* Sykes	Golden Oriole

Sub Family : Dicrurinae

Dicrurini

55. *Dicrurus macrocercus* Vieillot	Black Drongo
56. *D. paradiseus* Linneaus	GreaterRackettailed Drongo

Monarchini

57. *Terpsiphone paradisi* Linneaus	Asian Paradise Flycatcher

Sub Family : Aegithinae

58. *Aegithina tiphia* Gmelin	Common Iora

Family : Muscicapidae

Saxicolini

59. *Copsychus saularis* Sclater	Magpie Robin

Family : Sturnidae

60. *Sturnus m. malabaricus* Gmelin	Greyheaded Myna
61. *Accridoteres tristis* Linneaus	Common Myna
62. *A. fuscus* Sykes	Jungle Myna

Family : Paridae

Sub Family : Parinae

63. *Parus major* Hartert	Grey Tit

Family : Hirundinidae

64. *Hirundo rustica* Scopoli — Common Swallow
65. *H. daurica* Hodgson — Redrumped Swallow

Family : Pycnonotidae

66. *Pycnonotus cafer* — Redvented Bulbul

Family : Cisticolidae

67. *Cisticola juncidus* Whistler — Streaked Fantail Warbler
68. *Prinia inornata* Blyth — Plain Wren Warbler
69. *Prinia socialis*. Blyth

Family : Sylvidae

Sub Family : Accrocephalinae

70. *Orthotomus sutorius* Latham — Tailor Bird
71. *Accrocephalus dumetorum* Blyth — Blyth's Reed Warbler

Sub Family : Sylvinae

72. *Phylloscops trochiloides* Blyth — Greenish Leaf Warbler
73. *P. magnirostris* Blyth — Largebilled Leaf warbler

Timalini

74. *Turdoides striatus* Jerdon — Jungle Babbler

Family : Nectarinidae

Sub Family : Nectarinae

Dicaeini

75. *Dicaem erythrorhynchos* Latham — Tickle's Flowerpecker

Nectarini

76. *Nectarinia zeylanica* Hermann — Purplerumped Sunbird
77. *N. asiatica* Latham — Purple Sunbird
78. *N. lotenia* Whistler — Loten's Sunbird

Family : Passeridae

79. *Ploceus philippinus*. Linn — Baya Weaver

Sub Family : Mitacillinae

80. *Dendronanthus indicus* Gmelin — Forest Wagtail

81. *Motacilla maderaspatensis* Gmelin — Large Pied Wagtail

Sub Family : Estrildinae

82. *Lonchura striata* Linneaus — Whitebacked Munia

Table 1. Species diversity in Families

Sl.No	Families	Species
1.	Anatidae	2
2.	Picidae	1
3.	Megalaimidae	2
4.	Alcedinidae	1
5.	Halcyonidae	2
6.	Cerylidae	1
7.	Meropidae	2
8.	Cuculidae	1
9.	Centropopidae	1
10.	Psittacidae	1
11.	Apodidae	3
12.	Strigidae	1
13.	Columbidae	1
14.	Rallidae	3
15.	Scolopacidae	1
16.	Charadridae	1
17.	Jacanidae	1
18.	Laridae	2
19.	Accipitridae	4
20.	Anhingidae	1
21.	Phalacrocoracidae	3
22.	Ardeidae	9
23.	Threskiornithidae	2
24.	Podicipedidaea	1
25.	Ciconodae	1
26.	Pittidae	1

27.	Irenidae	1
28.	Corvidae	10
29.	Muscicappidae	1
30.	Sturnidae	3
31.	Paridae	1
32.	Hirundinidae	2
33.	Pycnonotidae	1
34.	Cisticolidae	4
35.	Sylvidae	5
36.	Necrtarinidae	4
37.	Passeridae	4

Table 2. Family Diversities in Orders

Sl. No.	Order	Families
1	Anseriformes	1
2	Piciformes	2
3	Coraciformes	4
4	Cuculiformes	2
5	Psittaciformes	1
6	Apodiformes	1
7	Strigiformes	1
8	Columbiformes	1
9	Gruiformes	2
10	Ciconiformes	9
11	Passeriformes	13

Table 3. List of Migratory Bird Species

Sl N.	Scientific Name	Common Name	Family
1	*Merops philippinus*	Bluetailed Bee-eater	Meropidae
2	*Actitis hypoleucus*	Common Sandpiper	Scolopacidae
3	*Gelochelidon nilotica*	Gullbilled Tern	Laridae
4	*Childonias hybridus*	Whiskered Tern	,,
5	*Circus aeruginosus*	Western Marsh Harrier	Accipitridae
6	*Plegadis falcinellus*	Glossy Ibis	Threskiornithidae

7	*Threskiornis melanocephalus*	White Ibis	,,
8	*Pitta brachyura*	Indian Pitta	Pittidae
9	*Oriolus oriolus*	Golden Oriole	Corvidae
10	*Terpsiphone paradisi*	Asian Paradise Flycatcher	,,
11	*Sturnus m. malabaricus*	Greyheaded Myna	Sturnidae
12	*Hirundo rustica*	Common Swallow	Hirundinidae
13	*Accrocephalus dumatorem*	Blyth's Reed Warbler	Sylvidae
14	*Phylloscops trochiloides*	Greenish Leaf Warbler	,,
15	*Phylloscops magnirostris*	Largebilled Leaf Warbler	,,
16	*Dendronanthus indicus*	Forest Wagtail	Passeridae

Table. 4. List of Breeding Birds

Sl No.	Scientific Name	Common Name	Family
1	*Megalaima viridis*	Small Green Barbet	Megalaimidae
2	*Psittacula krameri*	Roseringed Parakeet	Psittacidae
3	*Corvus splendens*	House Crow	Corvidae
4	*Parus major*	Grey Tit	Paridae
5	*Prinia inornata*	Plain Wren Warbler	Cisticolidae

References

A Brief Outline of Biodiversity Conservation and Action Plan for Kerala. 2003. Kerala Forest and Wild Life Dept.

Ali, S. and Ripley, S.D. 1968-78. *Hand Book of the Birds of India and Pakistan. Vol. I, II, III, IV, V, VI, VIII, IX and X.* Oxford University Press. New Delhi.

Ali, S. 1953. *The Birds of Kerala.* Oxford University Press. Bombay

Balasubramanian, P and Maheswaran, B. 2003. Frugivory, Seed Dispersal and Regeneration by Birds in South Indian Forests. *Jour. Bomb. Nat. Hist. Soci. Vol. 100 (2&3).*

Bates, R.S.P. and Lowther, E.H.N. 1952. *Breeding Birds of Kashmir.* Oxford University Press. New Delhi.

Elphick, J. (Gen. Ed.). 1995. Atlas of Bird Migration. Random House Inc. New York.

Gandhi, T. 2000. *Birds and Plant Regeneration.* Ravi Dayal Publisher. New Delhi.

Grewal, B., Harvey, B. and Psifter, O. 2002. *A photographic Guide to the Birds of India and the Indian Sub-continent including Pakistan, Nepal,*

Bhutan, Bangladesh, Sri Lanka and the Maldives. Periplus Edition Ltd. Singapore.

Grimmett, R., Inskipp, T. and Inskipp, C. 2000. *Pocket Guide to the Birds of the Indian Sub-continent*. Oxford University Press. Singapore.

Jayson, E.A. 2002. Avifauna in the Wetlands of Kerala. *Wetland Conservation and Management in Kerala*. State Committee on Science, Technology and Environment. Thiruvanathapuram.

Neelakantan, K.K. 1996. *Keralakthile Pakshikal*. Kerala Sahithya Academy, Thrissur (Malayalam).

Panickar, S.D. 1996. *Animal Diversity*. Zoological Society of Kerala. Kottayam.

Parker, T.J., Haswell, W.A and Marshall, A.J. 1898. *Text Book of Zoology. Vol. II: Vertebrate*. The Mac Millan Press Ltd., Hong Kong.

Rahmany, A.R. (Ed.) 2002. *Final Report: Asian Water Bird Census, India, 2002*. BNHS. Mumbai

Sreekumar, B. (Ed). 2003. *Vembanad Water Bird Count*. Dept. of Forest and Wild Life. Govt. of Kerala.

Unni, P.N. 2002. Biodiversity of Mangroves and avifauna in Wetland of Kerala: Need and strategies for Conservation. *International Conference on Biodiversity and Environment*. OISCA South India Chapter.

BIODIVERSITY RICH AREAS OF ORISSA: A NEED FOR CONSERVATION

C. Pattanaik, Sudhakar Reddy and M.S.R. Murthy

Forestry and Ecology Division, National Remote Sensing Agency, Hyderabad-500 037

Abstract

Sustainable management of natural resources has become a key issue for survival of life on earth. Conservation of biodiversity has been put to the highest priority through Convention on Biological Diversity (CBD). After a brief survey of the forests in Orissa, six potential sites highly rich in floristic composition and endemism have been identified. These are designated as the biodiversity granaries of Orissa. They are the Similipal Biosphere Reserve, Gandhamardan hill, Mahendragiri hill, Deomali hill, Bhitarkanika and Chilika lagoons. Out of the 2727 plant species reported from Orissa, 27 species have been found to be endemic and all most all of them have been found to be present in these granaries. The present paper deals and discusses the six bio-granaries along with their physiographical significance, diverse vegetation types and important floral elements. Ecological degradation and loss of biodiversity pose a serious threat to sustainable development. In recent times, indiscriminate human exploitation of natural resources and over population coupled with the continuous grazing of (un der) productive cattle population pose the greatest threat to species and ecosystems. Many valuable species have reduced to an alarming number and at times disappeared from the scene. Natural hazards owing to their adverse impacts, also contribute to the loss of biodiversity. Thus, an urgent need has been felt to conserve the gene pool involving a mix of *in situ* as well as *ex situ* strategies.

Introduction

India is one of the 12-mega centers of biodiversity with unique floristic composition. So far 17,500 species have been discovered in which 33% of the angiospermic species are endemic to India. With varied agro-climatic conditions, diversified edaphic factors, different altitudinal ranges from sea level to a few thousands of metres, changing rain patterns, Orissa has enormous genetic variation in natural flora. Orissa has extending over an area of 1, 55,842 sq.km on the east coast of India between 17 49' N to 22 34'N and 81 27'E to 87 29'E. It is surrounded by Jharkhand and West Bengal on north, on the west by Chhatisgarh, on the south by Andhra Pradesh and on the east by Bay of Bengal. Out of 2727 species are reported by *Saxsena & Brahmam*, 27 species are endemic to Orissa. Physiographically the entire Orissa is divided into 4 regions. These are-

(1) The Northern plateau

(2) The Central river basin

(3) The Eastern hills

(4) The Coastal plains

The Northern plateau is intersected by hill ranges gradually sloping down from north to south. The central river basin comprised Mahanadi, Brahmani, Baitarani rivers flowing from western and northern plateau to Bay of Bengal. The eastern hills as a part of Eastern Ghats extending from south-west to central Orissa contain some high peaks which harbour diversified taxa. The coastal plain extending from Balasore in the north to Berhampur in south is divided into two eco-regions. One for the mangrove vegetation at Bhitarkanika and other for the rich flora and fauna of Chilika Lake, the biggest brackish water lagoon of India. According to FSI report (2001) Orissa has now 58,135 sq. km of forest covers which is nearly 37% of geographic area.

Study area

Though Orissa harbours a diversified floristic composition, six forest pockets were found as richest granaries of biodiversity after studying the different vegetational patterns coupled with extensive field surveys. These are:

(1) Similipahar hills of Mayurbhanja district

(2) Gandhamardan hill ranges of Bolangir and Baragarh districts

(3) Mangroves of Bhitarkanika and Mahanadi delta

(4) Mahendragiri hill ranges of Gajapati district

(5) Deomali hill ranges of Koraput district

(6) Chilika lake and its environs

Each granary having plenty of diversified flora harbouring a few unique taxa specific to that locality and not found elsewhere in the state.

(1) Similipahar hills

Similipal lies between 86°10' E to 86°20' longitudes and 21°27' N to 22°01' N latitudes on the north eastern part of the state. Luxuriant forest cover seen in Similipahar National Park. It covers an area of 2750 sq. km. There are 1012 wild taxa and 64 cultivated species spread over 168 families (Saxena & Brahmam, 1989). The vegetation of Similipahar forest can be classified into 5 types as per Champion & Seth (1968). These are-

(1) Northern tropical moist deciduous forest

(2) Northern tropical semi-evergreen forest

(3) Dry deciduous hill forest

(4) Dry sal forest

(5) Grasslands

Many species have been found to be new records for the Botany and Orissa. These are - *Colysis pedunculate* (Hook. & Grev.) Ching, *Salomonia cantoniensis* Lour, *Oxalis lat*ifolia Kunth, *Ziziphus glabrata* Heyne ex Roth, *Cissus assamica* (Lawson) Craib, *Luffa cylindrical* (L.) Roem. var. *minor* chakravarthy, *Anaphalis adnata* DC, *Blumea aromatica* (Wall.) DC, *Blumea clarkei* Hook.f., *Embelia floribunda* Wall., *Cynanchum tunicatum* (Retz.) Alston, *Toxocarpus kleinii* W. & A, *Cordia wallichii* G. Don, *Aneilema ovalifolium* (Wight) Hook. f. & C.B. Clarke and *Rhapidophora glauca* schott. Some species are exclusively found in these area and not anywhere in the state. These are - *Dicrocephala integrifolia, Hypericum gaitti, Cyathia spinulata, Gomphostemma parviflorum, Neocinnamomum caudatum, Saphora bekeri, Toxocarpus kleinii* etc. Several tribals are living inside the similipahar forest areas and they are fully dependent on forest. Many ethnobotanical notings made by several persons depicts that the forest has a rich wealth of medicinal plants.

(2) Gandhamardan hill ranges

Gandhamardan hill ranges lie between 82°46' to 83°01' E longitudes and 20°48' to 20°55' N latitudes in Bolangir and Baragarh districts bordering Chhatisgarh state. It covers an area of nearly 300 sq. km having hill ranges between 250-1050 m in height. The luxuriant vegetation can be divided into 4 types. These are-

(1) Dry deciduous forest

(2) Semi-evergreen forest

(3) Bamboo formations

(4) Grasslands

There are as many as 781 species of flowering plants present in these hill ranges. Some interesting and specific species of Gandhamardan area are - *Erythrina resupinata, Tylophora fasciculatae, Heterostemma tanjorense, Actinoplaphne angustifolia, Pittosporum wightii, Protium serratum* etc.

(3) Mangroves of bhitarkanika and mahanadi delta

Bhitarkanika is only famous for mangroves in the east coast of India. Government of Orissa declared Bhitarkanika as a sanctuary in 1975. Later Government of India declared it as a National park in 1998. It lies between 86°49' E to 87°00' E longitudes and 20°40' to 20°43' N latitudes harbouring some rare and endangered mangrove species. It is formed due to inter-tidal estuarine complexes of Mahanadi, Dhamra, Brahmani, Baitarani and Devi rivers. The deltaic zone having high salinity, ph, high humidity, water logging and periodic tidal flushing creates a suitable environment for mangroves to grow vigorously. Out of 130 mangrove species (its associates) found in India, more than 80 species are found in Bhitarkanika and its adjoining areas. Some of the mangrove species found in Orissa are- *Avicennia marina, Rhizophora stylosa, Thespesia populnoides, Heritiera littoralis, Sonneratia alba, Sonneratia griffithii, Salvadora persica* etc. Due to high anthropogenic activity some mangroves are under constant threat. These are- *Ceriops decandra, Cerbera manghas, Rhizophora apiculata, Rhizophora mucronata, Kandelia kandel, Bruguiera cylindrical, B. parviflora, B. sexangula, Brownlia tersa, Avicennia marina, Avicennia officinalis, Tamarix troupii,*

Cynometra iripa, Phoenix paludosa, Heritiera fomes, Sonneratia apetala, Aegialites rotundifolia, Aegiceras corniculatus, Xylocarpus granatum and Xylocarpus gangeticus etc.

(4) Mahendragiri hill ranges

Mahendragiri hill ranges are important part of Eastern Ghats lie in Gajapati district. It is the second highest peak in orissa (1501 m). It has a diversified flora is found in these hill ranges due to subtropical climate favours some temperate species to invade. Some of the lower Himalayan and south Indian species are also found here. These are- *Maoutia puya, Viola serpens, Viola betonicifolia, Clematis roylei, Pittosporum nepalensis, Anotis calycina, Ajuga maerosperma, Thalictrum foliolosum, Rubus ellipticus, Sarcococca trinervia, Zanthoxylum armatum, Ophiopogon intermedius, Tephrosia roxburghiana, Sophora interupta, Litsea latea, Neolitsea zeylanica, Diosyros candolleana, Pavetta brevifolia, Weldenlandia gamblei* etc. Many species have been found to be new records for the Botany of Bihar and Orissa. These are- *Aganosma cymosa, Cyanotis vaga, Justicia glabra, Moutiya puya, Paspalum conjugatum, Plectranthus nilhericus, Psychotria fulva, Neanotis quadrilocularis, Ophiorhiza trichocarpa, Eulalia quadrinervis* etc. Three new species reported from this hills for the first time. These are - *Dimeria mahendragiriensis Ravi, Saxsena & Brahmam, Habenaria panigrahiana S.misra, Habenaria panigrahiana var. parviloba S.misra*. It always looks lush green in every season because several types of forest are there. Its floristic composition of this hill is unique and found nowhere in Orissa.

(5) Deomali hill ranges

Deomali hill range which is the highest peak in Eastern Ghats lies Koraput district. It is the tallest hill in Eastern Ghats as well as in Orissa having 1672 m height. The vegetation can be divided into four types.

(1) Moist deciduous forest

(2) Sub-tropical broad leaf hill forest

(3) Scrub savannah

(4) Grasslands

Some species are specific to this hill and not found elsewhere in the state. These are - *Habenaria grandifloriformis, Emilia zeylanica, Gynura lycopersicifolia* etc. Many of the tribals are settled down in these hilly areas. They are exploiting the plant resources for their food, household and medicinal purposes. The main Tribes which are (kondha and savara communities) residing in the interior forest areas, doing shifting cultivation. That automatically degraded the forest areas.

(6) Chilika lake and its environs

Chilika is the largest brackish water lagoon of India. It is situated in Puri, Khurda and Ganjam districts. Recently it has been awarded Ramsar sites of the world. Spreading over an area of 1100 sq. km the lake and its surrounding areas harbouring many hydrophytic flora. It attracts thousands of colourful birds every year from different countries particularly Siberian states. A number of islands like Nalabana, Bird's island, Malud, Parikuda, Ghantasila, Badakuda and Sanakuda etc. are present which expanded the lake. Nearly 350 sq.km flowering plants have been found in Chilika and its environment. Some of the important ones are - *Sueda maritima, Sesuvium portulacastrum, Salicornia brachata, Najas marina, Ruppia maritima, Cressa cretica, Halophila ovalis, Halophila beccarii, Azima tetracantha, Potamogeton octandrus, Potamogeton pectinatus, Najas graminea* etc. Really it is a natural paradise for the people as well as birds.

Conservation strategy

The luxuriant forest cover of Orissa has been devastated due to constant operation of biotic factors of various categories. Several anthropogenic activities are going on inside the forest, depleting the forest cover day by day. Shifting cultivation is also a major threat to forest depletion. Tribals are mercilessly cutting down the trees and cleared the forest areas for cultivation.

The steps needed to be taken for conservation:-

1 All potential biodiversity rich areas to be identified demarcated and mapped.

2 Degraded patches need to be restored and given adequate protection.

3 Shifting cultivation to be protected.

4 Joint Forest Management with adequate eco-development packages should be encouraged and practiced.

5 Awareness should be created among the people on the protective, productive and ecological role of forests.

6 Plantation programmes should go on regularly.

7 Research need be taken up on various aspects of forest species like propagation, hydrology, economics and ecology etc.

8 Strengthening livelihood security of the local community who are living adjoining to the forest areas.

9 Empowering the stakeholders in adopting participating Forest Management practices.

10 Forest protection act and Environment act should be strictly regulated.

Conclusion

In this alarming situation conservation of these biodiversity rich wealth is very important. Most of these plants have been reduced to a greater extent and even some endemic and endangered species disappeared from these areas of Orissa. Therefore, immediate steps must be taken in these areas for in-situ as well as ex-situ conservation of these taxa. This task should be undertaken not only by the government but also by the non-government organizations (NGO's). More forest areas should be protected by declaring them as sanctuaries, reserve forest and so on. This will facilitate to grow the endemic taxa in the natural environment. Use of natural forest areas for development work should not be allowed. Plants are conserved in Botanical gardens and seeds are also stored in seed banks. Propagation of rare and endangered plants should be made through seeds as well as through vegetative propagation.

However, with a better understanding of the concept and it's across the-broad applicability, the state of Orissa is reorienting its various programmes to fulfill the national and international commitments towards biodiversity conservation.

Acknowledgement

The authors are grateful to Dr. R.R. Navalgund, Director, National Remote Sensing Agency and Dr. P.S. Roy, Deputy Director (RS & GIS, Application Area), NRSA for their valuable suggestions and encouragement. We are also thankful to Department of Biotechnology (DBT) and Department of Space (DOS), Govt. of India, for providing financial assistance to carry out this work. The help received from colleagues of Forestry and Ecology Division, NRSA are thankfully acknowledged.

References

Anonymous, 1993. Mahendragiri - The pride of Eastern Ghats, pp-10, Orissa Environmental Society, Bhubaneshwar.

Anonymous, 2001. Strategy and Action plan on biodiversity of Eastern Ghats-ecoregional report for National Biodiversity Strategy and Action plan.

Champion, H.G. and Seth, S.K. 1968. Revised survey of forest types in India, Government of India, Press publication, New Delhi.

Gamble J.S. and Fischer, C.E.C. 1967. *Flora of Presidency of Madras*, Vol. I-III (repr. ed.), Botanical Survey of India, Kolkata.

Haines, H.H. 1926. *The Botany of Bihar and Orissa*, London.

Hajra, P.K. and Mudgal, V. 1997. Plant diversity hotspots in India, An overview, Botanical Survey of India, Kolkata.

Hooker, J.D. 1872-97. *The Flora of British India*, London.

Jain, S.K. and Sastry, A.R.K. 1980. Threatened plants of India: A state of the art report, Botanical Survey of India, Kolkata.

Mishra, M.K., Dash, S.S. and Das, P.K. 1999. Additions to the flora of Orissa, *Rheeda* 9: 163-172

Mooney, H.F. 1950. *Supplement to the botany of Bihar and Orissa*, Ranchi.

Nayar, M.P. 1977. Changing patterns of the Indian flora, *Bull. Bot. Surv. India* 19: 145-146

Nayar, M.P. and Sastry, A.R.K. 1987-90. *Red data books of Indian plants*, Vol I-III, Botanical Survey of India, Kolkata.

Panda, G.K. and Tripathy, S.N. 1993. Mahendragiri and its environs: Geographical overview, in Mahendragiri- The pride of Eastern Ghats, pp 7-15, Orissa Environmental Society, Bhubaneshwar.

Panigrahi, G., Choudhury, S., Raju, D.C.S. and Deka, G.K. 1964. A contribution to botany of Orissa, *Bull. Bot. Surv. India* 6: 237-266

Raju, D.C.S., Ahmedullah, M. and Nayar, M.P. 1987. Genetic potential in the flora of the Eastern Ghats of India, *Jour. Econ. Taxon. Bot.* 9(1): 133-138

Reddy, C.S., Reddy, K.N. and Jadhav, S.N. 2001. Threatened medicinal plants of Andhra Pradesh, Medicinal Plant Conservation Centre, FRLHT, Hyderabad.

Saxsena, H.O. and Brahmam, M. 1996. *The Flora of Orissa*, Vol I-IV, Orissa Forest Development Corporation, Bhubaneshwar.

ETHNO-MEDICO BOTANICAL EXPLORATION OF SOUTHERN ORISSA

C. Pattanaik, Sudhakar Reddy, N.K. Dhal[1] and Ashok Panigrahi[2]

National Remote Sensing Agency, Hyderabad-500 037
[1]Regional Research Laboratory, (C.S.I.R.), Bhubaneswar-751 013
[2]Botany Department, Berhampur University, Berhampur-760 007

Abstract

A major part of the population of India live in the rural areas and a considerable proportion of them are tribals, residing in the interior forests. It is imperative that a lot of ethnobotanical information in their folklore remains unnoticed and undocumented due to lack of communication. The different traditions, needs, culture of various tribes and the diversity of flora richly contribute to the Indian plant folklore. The state of Orissa is a rich repository of medicinal plant diversity. The southern part of Orissa constitutes of several hill-ranges belonging to the Eastern Ghats region. Orissa has 62 tribal communities. Many of them are restricted to interior hill forests only. The principal tribes seen in this region are Bonda, Kondh, Gond, Saura, Koya, Diyadi, Dharua, Paroja, Lodha, and Bhatoda etc. The tribals utilize many ethnoherbals from their surrounding vegetation for the treatment of various diseases and disorders. The present paper highlights some of the high valued ethno medicinal plants along with information on local names, locality, ethno medicinal uses, dosage and mode of administration, as practiced by them. Some of the important medicinal plants are *Achyranthus aspera, Alangium salvifolium, Aole barbadensis, Bauhinia vahili,*

Bombax ceiba, Holarrhena antidysentrica, Tridax procumbens, Calotropis gigantea, Rauwolfia serpentina, Caryota urens, Costus speciosus, Justicia adhatoda, Terminalia chebula, Terminalia bellirica etc. These valuable data gathered during the study on medicinal uses of various plants among the tribals are presented in this paper for further scientific exposure. Care should be taken to enlist the details of all tribal medicines and these should be patented, as this rich repository of traditional Vedic medicines can serve as a boon for the pharmaceutical industries.

Introduction

Ethnomedicines have their traditional base developed by cultural evolution in different parts of the world. All indigenous remedies have their roots in ethnomedicobotany. Along with the development of Ayurveda as a science of life and health three thousand years ago, studies have also been made in the identification of plants of medicinal values and also about their quality, uses and remedies for specific diseases (Dayal, 1977). Besides the written records of the uses of medicinal plants, some knowledge on the subject remains amongst the present day aboriginal societies and tribes and also with the villagers of remote areas which descended through generations of oral folklore. With the passage of time, many of these valuable information have been lost and attempts are now being made to collect the surviving bits of that traditional oral folklore. Traditional medicine, which includes ethnomedine (WHO, 1978) is important as it provides services to 75-80% of the world's population (Marine-Bettolo, 1980). Therefore, both traditional medicinal knowledge as well as the medicinal plants is important from the human health point of view that needs conservation.

Study area

Orissa is a rich repository of medicinal palnt diversity. Southern part of Orissa covers almost seven districts. They are- Malkangiri, Koraput, Nabarangpur, Rayagada, Phulbani, Kalahandi and Gajapati. These are inhabited by different tribes such as Kondha, Saura, Koya, Bonda, Gadaba, Bhatoda etc. Different tribes are found in each districts are given in Table 1.

The forest vegetation of these districts is of following types. (1) Semi evergreen forest, (2) Moist deciduous forest, (3) Dry deciduous forest. Most of the tribal people of these districts are living in the interior forest areas having no connection to the outward world.

Table 1. Different tribes found in each districts of Orissa

Sl No.	Districts	Tribes
1	Phulbani	Kondha
2	Kalahandi	Kondha, Lodha
3	Gajapati	Bhatoda, Saura
4	Rayagada	Kandha, Saura, Dangaria, Jhodia
5	Nabarangpur	Paraja, Gadava, Bhumia, Domb
6	Koraput	Gond, Kulis, Bonda, Koya, Gadava
7	Malkangiri	Bonda, Dadiya, Saura

They depend on traditional medicines for their various treatments. Although many of them are aware of the modern allopathic medicines, the economic conditions of these tribes do not permit them to undergo modern treatment. Very little work has been done on the ethno-medico-botany of south Orissa, though sporadic reports on this aspect are available (Choudhury *et al.*, 1975; Patnaik *et al.*, 1986; Das & Mishra, 1987; Dash & Mishra, 1996). In this paper an attempt has been made to collect all the information available on ethno medicinal uses of plants by different tribes of South Orissa.

Materials and Methodology

The present paper is the outcome of the study done during 2001-2003. Several field trips were made to different areas of south Orissa covering all the seasons during the study. The economic and ethno botanical data of the species were collected from the Baidas, Majhis and Headman of the concerned villages. Besides, notes on habit, habitat, flowering period and distribution of species in specific areas were also noted. A comprehensive data on plant names, its local names, family, plant parts used, locality where it found and its uses in curing different diseases are given in Table 2.

Table 2. Name of the plants, parts used and the diseases against which plants are used by the tribes of South Orissa

SN.	Name of the plants	Local name	Family	Parts used	Locality	Diseases
1	*Abrus precatorius* L.	Kaincha	Fabaceae	Root	Potangi	Abortifacient
2	*Acalypha indica* L.	Indram-aricha	Euphorbi-aceae	Whole plant	Muniguda	Laxative, Mouth infection

3	*Achyranthes aspera* L.	Apamar-anga	Amaran-thaceae			Toothache, quick delivery
4	*Aegle marmelos* (L.) Corr.	Bela	Rutaceae	Leaf, Fruit	Phulbani	Dieabetes, Dysentery
5	*Aerva lanata* (L.) Juss.ex Schultes	Paunsia	Amarant-haceae	Whole plant	Chitra-konda	Fever
6	*Ageratum conyzoides* L.	Pokasunga	Asteraceae	Whole plant	Chandr-apur	Diarrhoea
7	*Alangium salvifolium* Lam.	Ankula	Alangiaceae	Stem bark	Kalimela	Conjuctivitis
8	*Aloe barbadensis*	Ghikuari	Agavaceae	Leaf	Muniguda	Headache
9	*Andrographis panic-ulata* (N.Burm.) Wall	Bhuin nimba	Acanthaceae	Leaf	Balimela	Itches, Skin diseases
10	*Ardisia solanacea* Roxb.	Tinkoli	Myrsinaceae	Fruit	Chitra-konda	Headache
11	*Asparagus recemosus* Willd.	Satabari	Liliaceae	Root	Gupteswar	Check excess menstrual bleeding
12	*Azadirachta indica* A. Juss.	Nimba	Meliaceae	Stem bark	Jeypore	Dermatitis, Diabetes
13	*Barleria montana*		Euphorbi-aceae	Leaf	Sileru road	Headache
14	*Boerhaavia diffusa* L.	Puruni	Nyctagin-aceae	Leaf	Laxmipur	Diuretic
15	*Bombax ceiba* L.	Simuli	Bombacaceae	Root	Malkangiri	Abscess bursting
16	*Buchnania lanzan* Sprang.	Char	Anacardi-aceae	Stem bark	Umerkote	Applied in burn areas
17	*Butea monosperma* (Lam.) Taub.	Palasha	Fabaceae	Root	Phulbani	Contrac-eptive
18	*Calotropis gigantea* (L.) R.Br.ex Aiton	Arakha	Asclepia-daceae	Leaf, Latex	Every where	Snakebite, Sciatica
19	*Cassia fistula* L.	Sunari	Caesalpi-niaceae	Stem, Bark	Kotpada	Laxative
20	*Cassia occidentalis* L.	Chakunda	Caesalp-iniaceae	Root	Roadside	Abdominal di sorder, Fever
21	*Celastrus paniculatus* Willd.	Kharasana	Celastraceae	Seed, Root	Govind-apalli	Eczema, Abortifacient
22	*Cipadessa baccifera* (Roth.) Miq.	Nahalabeli	Meliaceae	Stembark	Chitra-konda	Vitiated bile

23	*Cissus quadrangularis* L.	Hadab-hanga	Vitaceae	Whole plant	Jharigaon	Jointpain, Bone-setting
24	*Cleistanthus collinus* (Roxb.) Benth.	Karada	Euphorb-iaceae	Leaf	Kudum-ulgumma	Arborti-facient
25	*Costus speciosus* (Koenig) Sm.	Kudha	Zingiberaceae	Rhizome	Koraput	Epilepsy, Sperma-torrhoea
26	*Curculigo orchioides* Gaertn.	Talamuli	Hypoxidaceae	Rhizome	Chitra-konda	Healing wounds, Gonorrhoea
27	*Curcuma longa* L.	Haladi	Zingiberaceae	Rhizome	Phulbani	Affected areas of boils
28	*Datura metel* L.	Dudura	Solanaceae	Root	Jeypore	Pimples, Boils
29	*Dioscorea pentaphylla* L.	Bana alu	Dioscor-eaceae	Tuber	Baliguda	Sterility in males
30	*Elephantopus scaber* L.	Totamula	Asteraceae	Root	Darin-gibadi	Burns, Dysuria
31	*Erythrina indica* Lam.	Paladhua	Fabaceae	Stembark	Ramgiri	Menorrohea
32	*Euphrobia hirta* L.	Chitakuti	Euphorb-iaceae	Latex, Leaf	Every where	Conjuctivitis, Dysentery
33	*Ficus hispida* L.	Dimiri	Moraceae	Leaf	Chitra-konda	Conjuctivitis
34	*Gmelina arborea* Roxb.	Gambhari	Verbenaceae	Stembark	Gupteswar	Rheumatic pain, Fever
35	*Hemidesmus indicus* (L.) R.Br.	Ananta mula	Periplocaceae	Root	Every where	Dysentery, Rheumatism
36	*Holarrhena pubescens* (Buch.-Ham.) G.Don		Apocynaceae	Root, Latex	Kalimela	Diarrhoea, Toothache
37	*Imperata cylindrica* (L.) P.Beaur.	Chhanghas	Poaceae	Leaf	Podia	Protect bleeding from wounds
38	*Justicia adhatoda* L.	Basanga	Acanthaceae	Rootbark	Jeypore	Cough, Cold
39	*Kalanchoe pinnata* (Lam.) Pers	Amarpoi	Crassulaceae	Leaf	Malkangiri	Dysentery
40	*Lannea coromandelica* (Houtt.) Merill.	Mahi	Anacardi-aceae	Stembark	Balimela	Dysentery
41	*Madhuca indica* (Koenig.) Macb.	Mahula	Sapotaceae	Stembark	Koraput	Stomach pain

42	*Mallotus philippensis* (Lamk.) Muell-Arg	Sinduri	Euphorbi-aceae	Root	Kotapada	Earache
43	*Milletia extensa* Baker.	Gurendi	Fabaceae	Stembark	Rayagada	Fish poison
44	*Moringa oleifera* Lamk.	Sajana	Moringaceae	Stembark	Every where	Conjuctivitis, Fever
45	*Mucuna puriens* (L.) DC	Baidanka	Fabaceae	Root	Podia	Conception
46	*Nyctanthes arbortristis* L.	Ganga-siuli	Nyctanth-aceae	Leaf	Phulbani	Malaria Fever
47	*Oroxylum indicum* (L.) Vent	Pham-pani	Bignoniaceae	Stembark	Muniguda	Liver disorder, Backache
48	*Pergularia daemia* (Forst.) Chior.	Uturuli	Asclepia-daceae	Leaf, Latex	Panimunda	Malaria fev-er, Heeling wounds
49	*Phoenix acaulis* Buch-Ham ex Roxb.	Kojiri	Aerecaceae	Apical bud	Deomali hill	Puerperal disease
50	*Phyllanthus emblica* L.	Amla	Euphorbia-ceae	Leaf	Sileru road	Diarrhoea
51	*Plumbago rosea* L.	Rakta-chita	Plumbagi-naceae	Root	Baliguda	Abortifacient, Hydrocele
52	*Plumbago zeylanica* L.	Chita-paru	Plumbagi-naceae	Root	Govinda-palli	Paralysis, Snakebite
53	*Portulaca oleracea* L.	Puru-nisaga	Portulaca-ceae	Whole plant	Chitra-konda	Hookworm
54	*Pterocarpus marsupium* Roxb.	Bija	Fabaceae	Stembark	Ramgiri	Bool dysentery
55	*Rauvolfia serpentina* Benth.ex Kurz	Patala-garuda	Apocynaceae	Root	Boipari-guda	Snakebite, Menstrual pain
56	*Sida cordifolia* L.	Bisiripi	Malvaceae	Whole plant	Every where	Fracture
57	*Solanum surattense* Burm.f.	Anka-ranti	Solanaceae	Fruit	Malkangiri	Toothache, Asthama
58	*Soymida febrifuga* A. Juss.	Soma	Meliaceae	Stembark	Chitra-konda	Snakebite, Dysmeno-rrhoea
59	*Sterculia urens* Roxb.	Kodala	Sterculiaceae	Gum	Chitra-konda	Aphrodiasiac, Dysentery
60	*Riccinus communis* L.	Jada	Euphorbi-aceae	Leaf	Phulbani	Piles
61	*Rubia cordifolia* L.	Kata-singi	Rubiaceae	Root	Kotpada	Fever

62	*Schleichera oleosa* (Lour.) Oken	Kusuma	Sapindaceae	Stembark	Sileru road	Fracture
63	*Strychnos nux-vomica* L.	Kochila	Loganiaceae	Stembark	Kalinga	Skin disease, Contrace-ptive
64	*Syzigium cumini* (L.) Skeels	Jamu	Myrtaceae	Stembark	Jeypore	Diarrhoea
65	*Tamarindus indicus* L.	Tentuli	Caesalp-iniaceae	Leaf	Similiguda	Injuries
66	*Tephrosia purpuerea* (L.) Pers	Bana-nila	Fabaceae	Root	Nabara-ngpur	Abdominal pain
67	*Terminalia bellirica* (Gaertn.) Roxb.	Bahada	Combre-taceae	Seed, Fruit	Chitra-konda	Constipation, Cough
68	*Terminalia chebula* Retz.	Harida	Combre-taceae	Fruit	Chitra-konda	Purification of blood
69	*Tinospora cordifolia* (Willd) Hook.	Guluc-hilata	Meninsp-ermaceae	Stem	Govinda-palli	Cold, Jaundice, Snakebite
70	*Toddalia asiatica* (L.) Lam.	Tunda-poda	Rutaceae	Rootbark	Khairput	Epilepsy
71	*Tridex procumbens* L.	Bisal - yakarani	Asteraceae	Whole plant	Every where	Freshcuts, ch eck bleeding
72	*Urginea indica* (Roxb.) Hook.f.	Ban piajo	Liliaceae	Root	Balimela	Abscess on throat
73	*Vanda tessellata* Hook.f.ex G.Don.	Malanga	Orchidaceae	Leaf	Balimela	Otisis, Rheumatic pains
74	*Vernonia cinera* (L.) Less	Poka-sunga	Asteraceae	Root	Phiringia	Elephantiasis
75	*Vitex negundo* L.	Begunia	Verbenaceae	Leaf	Rayagada	Constipation, Stomatitis
76	*Woodfordia fruticosa* (L.) Kurz.	Dhataki	Lythraceae	Root	Ramgiri	Leucorrhoea, Pregnancy termination
77	*Zizyphus mauritiana* Lamk.	Barakoli	Rahman-aceae	Stembark	Jeypore	Abdominal pain, Headache

Results and Discussions

Forty-six angiospermic families represent these 77 medicinal plants under use. A maximum of 7 species are represented by each of Euphorbiaceae and Fabaceae; 4 by Asteraceae; 3 each by Caesalp-iniaceae and Meliaceae; 2 each by Acanthaceae, Amaranthaceae,

Anacardiaceae, Apocynaceae, Asclepiadaceae, Combretaceae, Liliaceae, Plumbaginaceae, Rutaceae, Solanaceae, verbenaceae and Zingiberaceae; 1 each by rest of the families.

Recently the traditional medicines has been proved very effective and curing various diseases. Scientific analysis also proves their medicinal properties which prescribed by these traditional healers. Those parts of the plants where the active compounds are deposited are important. In plants, these compounds are mostly secondary metabolites (alkaloids, steroids, flavonoids, tannins and saponins etc.) which are synthesized and deposited in specific parts or in all parts of the plant. These compounds are more complex and specific, and are found in certain taxa such as family, genus and species (Balandrin *et al*, 1988). It may be anticipated that random screening of plants for active chemicals is as important as the screening of ethno botanically targeted species (Farnsworth, 1988).

Therefore, conservation of medicinal plants is very important, as most of these plants have been reduced to a greater extent. Immediate steps must be taken in these areas for ex-situ as well as in-situ conservation of these taxa.

Acknowledgement

The authors are grateful to Dr. R.R. Navalgund, Director, National Remote Sensing Agency and Dr. P.S. Roy, Deputy Director (RS & GIS, Application Area), NRSA for their valuable suggestions and encouragement. We are also thankful to Regional Research Laboratory (C.S.I.R.) for verifying herbarium specimens. The help received from colleagues of Forestry and Ecology Division, NRSA are thankfully acknowledged.

References

Balandrin, M.; Klocky, J., Wurtele, E. and Bolinger, W. 1985. Natural plants chemicals: sources of industrial and medicinal materials, *Science*, 288: 1154-1160.

Choudhuri, Rai, H.N., Pal, D.C. and Tarfdar, C.R. 1975. Less known uses of some plants from the tribal areas of Orissa, *Bull. Bot Surv. India*, 17:132-136

Das, P.K. and Mishra, M.K. 1987. Some medicinal plants used by the tribal sod Deomali and adjacent areas of Koraput district, Orissa, *Indian Jour. Forester*, 10: 301-303

Dash, S.S. and Mishra, M.K. 1996. Tribal uses of plants from Narayanapatna region of Koraput district, Orissa, *Anc. Sci. life*, 15: 230-237.

Dayal, Rameshwal 1977. Chemical investigation of medicinal plants, *Indian forester*, 249-253.

Fransworth, N.R. 1988. Screening for new medicine pp 83-97, In: *Biodiversity* Wilson, E.O (Ed.) National Academic Press, London.

Marine-Bettolo, G.B. 1980. Presest aspects of the uses of plants in traditional medicine, *Jour. Ethnopharmacology* 2: 5-7.

Patnaik, N., Bara, F. and Mall, A. 1986. Herbal medicines of dongrais, *Adibasi*, 26: 1-4.

Saxsena, H.O. and Dutta, P.K. 1975. Studies on the ethnobotany of Orissa, *Bull. Bot. Surv. India*, 17: 124-131.

Saxsena, H.O. and Brahmam, M. 1996. *The flora of Orissa*, Vol I-IV, Orissa Forest Development Corporation, Bhubaneswar.

World Health Organisation 1978. The promotion and development of traditional medicine, *Technical report series 662*, Geneva.

EFFECTS OF LEAD ON WATER POLLUTION

Shibu P. Varghese, Anu Achuthan, Raimol Varghese and Sija Thankachen

Department of Botany, Mar Thoma College, Tiruvalla

Abstract

Rapid urbanization has led to pollution of water resources by various semi metals like lead. This paper investigates sources of lead pollution in water. It also focuses on the risk to human health associated with exposure to lead.

Introduction

Water is one of the most essential constituents of both flora and fauna. Man needs it in the first place for his physiological existence and many other purposes such as industrial water supply, irrigation, power generation and aquatic life etc. The ever increasing release of substances and energy as waste products of human activities results in harmful changes within the natural environment which has been termed pollution. As a consequence of rapidly expanding industrialization and excessive population growths, most of our rivers, lakes, streams and other water bodies are being increasingly polluted. Water is regarded as "polluted" when it is changed in its quality or composition, directly or indirectly as a result of mankind's activities so that it becomes less suitable for drinking, domestic, agricultural, recreational, fisheries or other purposes for which it would otherwise be quite suitable in its natural (unpolluted state). Public health authorities are now becoming greatly concerned about various toxic chemical pollutants found in waters. Current estimate indicate that about 75% of India's available water is polluted. The source of water pollution are sewage and other oxygen demanding wastes, agricultural wastes, physical pollutants and industrial wastes including various heavy metals.

Heavy metals are presently used in making different types of products like paints, dyes, electroplating, toys etc. Industrial effluents, volcanic eruptions, municipal sewage or refining, coal burning, cement production, fertilizers, vehicles, refineries, smelters, used batteries, power plants and piggeries etc., contribute different heavy metals to the environment. These metals can be buried on land fills or washed into sediments. After getting entry into aquatic bodies, these metals may precipitate or adsorb on solid surfaces, remain soluble or suspend in water or may be taken up by fauna and flora.

Degradation of organic materials in water results in the change of Eh and pH of water making the metal compounds more soluble and ultimately more available to aquatic organisms. Heavy metals are non-degradable unlike many organic pollutants and subsequently endanger flora and fauna. One of the major heavy metal pollutant is lead.

Lead has been in use for the last 6000 years. The Greeks and Romans used it for the manufacture of cooking utensils and in plumbing systems. Cragg (1968) wrote that millions of tons of lead were produced by the Greeks as a byproduct of silver refining. Greek and Roman aristocracy used lead-lined cooking, eating and drinking vessels. Lead salts gave a sweet taste to foods and retarded decomposition. Grape syrup was concentrated in lead lined pots over a fire and stirred to give the mixture a sweet taste. And thus it was that lead poisoning, with all its various effects, lassitude and infertility to mention only two, no doubt contributed to the collapse of the Roman Empire. Currently lead is being used widely:

Importance of Lead

1 Lead due to its ductility and higher resistance to erosion is widely used in many industries.

2 Use of tetraethyl lead and tetramethyl lead as an antiknock agent in industries is well known.

3 The lead used as stabilizer in some plastic pipes may be another source of pollution, especially for children.

4 Lead pollution through lead paints are greater hazards to children who are prone to ingest and chew on painted articles as lead toys; painters may also run a risk from continual use and exposure.

5 Lead used in insecticides, food beverages, ointments and various medicinal concoctions for flavouring and sweetening is also an important source of lead poisoning.

6 Lead is widely distributed in nature. It is usually associated with other metals, particularly silver and zinc.

7 The unique properties of lead-softness, malleability, low melting point, and resistance to corrosion make it one of the most widely used metals.

8 Inorganic lead compounds are widely used in the petroleum industry, the manufacture of storage batteries and the paint and pigment industries.

9 Large amounts of lead are used as solder in the soldering of seams and closures in the sheathing of telegraph, telephone and power cables.

10 A significant proportion of lead is used in the building industry in the form of sheets and pipes.

11 The ammonium industry is an important consumer of lead especially as lead-tin-antimony alloys

12 Lead is used in eye cosmetics, particularly in Asian countries.

How Lead Pollutes Water?

Lead (Pb) is the natural pollutant of water. Contamination of water bodies by lead appears to be mainly due to various human activities and improper waste management. Lead is accumulated in the marine environment as a result of leaching from rocks bearing this metal, industrial and domestic sources particularly the use of anti-knock motor fuels containing lead compounds. These include dumping of electric storage batteries, washing of vehicles and other machinery part, disposal of paint containers etc. The use of tetraethyl and tetra-methyl lead in petroleum industry may substantially contribute to the level of lead in the surrounding water bodies.

Lead enters in the aquatic environment through precipitation, lead dust fall out, erosion and leaching of soil, municipal and industrial waste discharge and the runoff and fall out deposits from streets and other surfaces. Lead is removed from the atmosphere by rain and falls back on the earth's surface and seeps into the ground.

Lead passes from soil to water and to the plants and finally into food chain.

The lead concentration in both sea water and fresh water has gone up. The concentration of Pb in sea water variesfrom 0.55 ppb to 86.7 ppb (Gupta and Salunkhe, 1985).

Lead Metabolism

1. Absorption

Lead may be absorbed into the body by ingestion (Gulson *et al.*, 1997), inhalation (Hodgkins *et al.*, 1991) or through skin (Stauber *et al.*, 1994). The absorption of lead from different sources is dependent on many factors, such as age and physiological status. For most individuals, the major route of absorption is the gastro-intestinal tract. Intestinal adsorption of lead by children is much greater than in the adults. Lead absorption through the skin is different from gastro-intestinal lead absorption with lead transported first into the plasma and rapidly concentrated into the extra cellular fluid pool of sweat and saliva without significant uptake by erythrocytes.

2. Distribution and Retention

Blood contains lead as a non-diffusible form bound to erythrocytes and as a diffusible form in plasma. Lead enters the blood stream where it rapidly becomes attached to red blood cells or is transported to soft tissues. Thus it is transferred to bone where the reservoir is large with a long half-life. Lead within RBC is bound primarily to haemoglobin (Ong and Lee, 1980). Lead in blood is primarily in RBC (99%) rather than plasma (Cake *et al.*, 1991; De Silva, 1981). Within the cells, 50% is bound to haemoglobin and other to proteins.

Among body organs the greatest percentage of lead is taken into the kidney, followed by the liver and other soft tissues such as heart and brain, but the lead in skeleton represents the major body fraction. Within the skeleton, lead is incorporated to the mineral in the place of calcium.

3. Excretion

Absorbed lead is eliminated primarily via the kidney in the urine, about 76% (Kehoe, 1987) and to a lesser extent by the gastro intestinal tract about 16% through billiary secretions (Conrad and Barton,

1978). Other routes for elimination (hair, nails, sweat and exfoliated skin) account for about 8%.

Biochemical and toxicological effects

1. Haematopoietic System

The haematological effects of led is due to the inhibition of haemoglobin synthesis, and shortened life span of circulating erythrocytes. These effects may result in anaemia, which may be mild, hyper, chronic and sometimes microcytic anaemia. Anaemia is also associated with shortened red blood cells life span. Basophilic stippling of RBC is a feature of lead induced anaemia.

The degree of shortening of erythrocytes correlates with levels of anaemia and blood lead. The biochemical basis of its toxicity is that it affects haeme biosynthesis, alters mitochondrial membrane structure and functions, impairs enzyme activities and damages brain cells.

2. Renal effects

Overt effects of lead on the kidney in man and experimental animals, particularly the rat and mouse, begin with acute morphological changes consisting of nuclear inclusion bodies or lead proteins complexes and ultra structural changes in organelles, particularly mitochondria. Long term exposure of lead may give rise to the development of irreversible functional and morphological renal changes including interstitial fibrosis, tubular atrophy and dilation.

It is still controversial whether exposure to lead in childhood is associated with the occurrence of nephropathy in the later life.

3. Neurological Effects

The effect of lead on cognitive and behavioural development of the central nervous system is the critical effect on infants and children. The generally recognized effect of lead on the central nervous system was encephalopathy. It was characterized clinically by ataxia, coma, and convulsions and was often fatal. Survivals suffered a number of neurological complications such as mental retardation, deafness, blindness and convulsions. A chronic form of encephalopathy has also been described (Al-Saleh, 1994) in which progressive mental retardation, loss of motor skills and behavioural disorders occur rather than the more precipitious symptoms seen in acute encephalopathy.

The primary effect of lead on the peripheral nervous system is reduced motor activity, muscular weakness, especially of the exterior muscles, tremor fatigue and lack of muscular coordination. One common feature of lead neuropathy and one of the characteristic feature of this disorder in adults is wrist drop to paralysis of the distal, upper extensor muscles.

It is suggested that the immature endothelial cells forming the capillaries of the developing brain are less resistant to the effect of lead than capillaries from mature brains. These cells permit fluids and cations including lead to read newly formed components of the brain, particularly astrocytes and neurons (Goldstein, 1990). The development of resistance to lead encephalopathy is thought by Holtzman *et al.*, (1984) to be related to the formation of lead-protein complexes in astrocytes and sequester lead away from mitochondria.

4. Reproductive Toxicity

Lead exerts its toxic effects directly on the developing foetus after gestation begins, and also indirectly on paternal or maternal physiology before and during the reproduction process. Several studies suggest that lead might increase the incidents of early membrane rupture and premature delivery and malformed births (Angel and Lavery, 1982). Increase in blood lead reduce semen volume, semen density and counts of total, motile and viable sperm, and percentage of pathological sperm.

Effects on plants

Plants are capable of absorbing heavy metals. The absorbed heavy metals can be discharged from leaf surface to atmosphere. Heavy metals are easily absorbed by the root and they travel all through the plant, and ultimately some of it is passed out through the leaf surface. The industrial wastes containing heavy metals pollute the water of the rivers, lake and streams. This water is used for irrigation purposes. The land become mineralized and some of this is absorbed by plants through their roots. These can also enter the human bodies through fruits and seeds of vegetable and fruit crop.

Acknowledgement

We express our whole hearted gratitude to Dr. Mary Chacko, Department of Botany, Mar Thoma College, Tiruvalla for her guidance, encouragement and suggestions throughout the period of the preparation of this review. We also thank Dr. Hari N. Iyer of the same department for his valuable suggestions and help.

We are grateful to Prof. Thomas Mathew, the Head and all other teaching staff of the Department of Botany, Mar Thoma College, Tiruvalla for providing all required facilitation during the course of our work.

We are indebted to Mr. Sajan Varghese, Librarian, Mar Thoma College, Tiruvalla for helping us access the internet.

References

Al-saleh, IAS. 1994. The Biochemical and Clinical consequences of lead poisoning *Med. Res. Rev.* 14 : 415-486.

Angel, N.F. and J.P. Lavery. 1982. The relationship of blood levels to obstetric outcome. *Am. J. Obstet. Gynaecol.* 142 : 40.

Barrey, P.S.I. 1981. Concentration of lead in tissue of children. *Br. J. Ind. Med.* 35 : 612-671.

Cake, K.M., R.J. Bowins, and C. Vaillancourt 1996. Partition of circulating lead between serum and red blood cells is different for internal and external sources of exposure. *Am. J. Ind. Med.* 29 : 440 – 445.

Conrad, M.R and J.C. Barton 1978. Factors affecting the absorption and excretion of lead in the rat. *Gastroenterology*, 74 : 731 –740.

Cragg, J.B. 1968. The Theory and Practice of Conservation, *IUCN Publication*, New Series, No. 12, 25-35.

De Silva, P.E. 1981. Determination of lead in plasma and the studies on its relationship to lead in erythrocytes. *Br. J. Ind. Med.* 38 : 209 – 217.

Goldstein, G.W. 1990. Lead poisoning and brain cell function. *Environ. Health. Perspect.* 89 : 91-94.

Gulson, B.L., C.W. Jameson and K.R. Mahaffey 1997. Pregnancy increased mobilization of lead from maternal skeleton, *J. Lab. Clin. Med.* 130 : 51-62.

Gupta, P.K. and D.K. Salunke 1985. *Modern Toxicology*, Vol. I-III. Metropolitan Books, New Delhi.

Hodgkins, D.G., T. G. Rogins and D.L. Hinkamp 1991. The effect of airborne lead particle size on worker blood-lead levels: An empirical study of battery workers. *Br. J. Ind. Med.* 49 : 241-248.

Holtzman P., C. Devries, H. Ngayan, J. Olson and K. Bensh 1984. Maturation of resistance of lead encephalopathy. Cellular and subcellular mechanisms. *Neurotoxocology,* 5 : 97-124.

Kehoe, R.A. 1961 Ecology. *J. Inst. Pub. Health*. Hyg. 24 : 24-177.

Kumar, H.D. 1996. Modern Concepts of Ecology, Vikas Publishing House, Pvt. Ltd.

Ong, C.N. and W.R. Lee 1980. Interaction of Calcium and lead in human erythrocytes. *Br. J. Ind. Med.* 37 : 78-84.

Shukla, R.S., P.S. Chandel 1972. Plant Ecology and Soil Science, S. Chand and Co Ltd., Ram Nagar, New Delhi.

Stauber, J.L., T.M. Florence and B.L. Gulson 1994. Percutaneous absorption of inorganic lead compounds. *Sci. Total Environ.* 145 : 55-70.

MEDICINAL VALUES OF *Catharanthus roseus* (L.) G. Don. IN ALKALOID PRODUCTION

Sarat Chandran A., Divakaran, Ginu Varghese, Liby Abraham and L. Seema

Department of Botany, Mar Thoma College, Thiruvalla

Abstract

Catharanthus roseus L. is a 'magic plant' that produces certain chemical compounds that are medicinally valuable. It is used in the treatment of childhood leukemia, Hodgkin's diseases etc. This paper is a review of the general medicinal properties of vincristine and vinblastine.

Introduction

A widely growing evergreen shrub, Periwinkle, has had the reputation of being a 'magic plant', for its beautiful flowers and leaves endowed with unique chemical compounds that have a high medicinal value. True to its name, derived from the Latin word 'Vincere' that means 'to overcome'.

They grow about two feet high and have glossy dark green leaves and flowers with colours ranging from white, pink to purple.

Catharanthus roseus L. formerly known by various names such as Madagascar Periwinkle, *Vinca rosea* L. belong to the family Apocynaceae, genus Catharanthus. In Hindi it is named *Sadabahar* and in Malayalam it is named as *Vishamalari* and in Tamil it is named as *Sudukadu mallika.*

Medicinal Importance

Of the three different Periwinkles, Madagascar or "Rosy" Periwinkle is by far the most important plant medicinally, generating hundreds of millions per year in the pharmaceutical industry as a promising addition to the arsenal against childhood leukemia, Hodgkin's Disease, testicular cancer and cancerous tumours, high blood pressure, and uncontrolled bleeding, not to mention the fact that it has powerful tranquilizing properties.

Historically, Rosy Periwinkle has been used to treat high blood sugar in diabetics, and many cultures still drink the herbal tea made out of the plant for this purpose. Unfortunately, recent testing has indicated that it has no effect on this particular disease.

Externally, Rosy Periwinkle leaves can be bruised and applied directly to the affected area for wasp stings, to relieve eye inflammation, and to stop bleeding. Alternatively, the leaves can be boiled and a cloth soaked with the liquid and applied to the affected area, or an herb ointment can be made using herb ointment from the plant.

Vinca major and Vinca minor appear to be quite similar as far as medicinal qualities, are concerned and internally are used in herb tea for improved oxygen and blood, especially in cerebral veins and arteries for improvement of cognitive function, tinnitus, depression, fatigue, problems associated with menopause, heavy menstruation, mild hypertension, headache, diarrhoea, congestion, as a gargle for sore throats, and for improving memory.

Externally, a herb ointment can be made, or the leaves can be bruised and applied directly for dermatitis, eczema, acne, bleeding gums, nosebleeds, mouth ulcers, and bleeding hemorrhoids.

Internal use

Periwinkle has been used successfully to treat leukemia in children. Verblastine sulfate is showing promising signs for use in Hodgkin's Disease and choriocarcinoma. It is also being used in lung cancer research. It helps stop constipation, internal hemorrhaging, bloody nose and bleeding piles. Periwinkle Tea is alone used to reduce nervousness, nightmares, hysteria and fits.

External use

A hair rinse made from Periwinkle Tea can help remove dandruff. A body wash of the tea can help skin sores, and other skin diseases.

Active compounds

There are two classes of active compounds in Vinca – alkaloids and tannins. The tannins present in Periwinkle have a mild astringent effect, which explains the traditional use of this plant in relieving pain from canker, sores and sore throats.

It contains over 100 alkaloids found in different parts of the plant. The important alkaloids that have been discovered in the aerial parts of Periwinkle include:

1 Catharathine
2 Leurosine Sulphate
3 Lochnerine
4 Tetra hydro alstonine
5 Vincamine
6 Vindoline
7 Vindolinine
8 Reserpine
9 Vincristine
10 Vinblastine
11 Ajamalicine
12 Serpentine

The main anti-tumour alkaloids extracted from Catharanthus include vincristine and vinblastine. Vincristine and vinblastine, in particular are bis-indole alkaloids.

These are minor alkaloids and their yields are typically 1 gm and 20 mg/1000 kg, respectively. Therefore, the isolation from leaves is a very tedious job and their rapid and accurate isolation is a field of interest for investigations. In general, the method used so far to obtain the alkaloid is solvent extraction by partitioning the alkaloid bases and their salts between organic / aqueous phases.

The leaves represent only about 10 per cent of the total dry weight of the plant, so the Chinese usually harvest nearly two million pounds of fresh plants at a time. Vinca minor, the common Periwinkle, has been grown for centuries in gardens. It yields an alkaloid known as vincamine, one currently under clinical investigation in France and elsewhere.

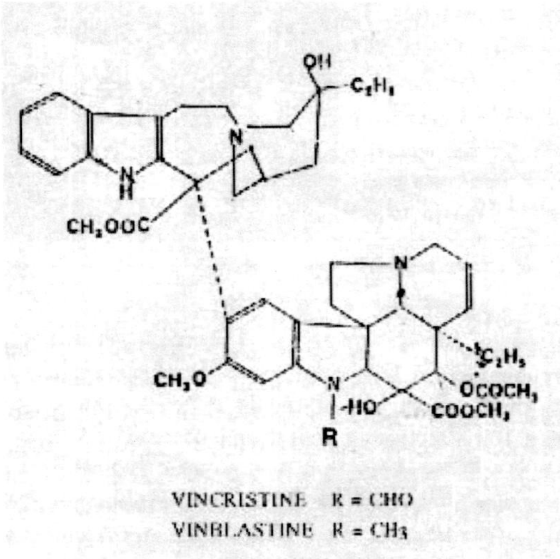

Vincristine and Vinblastine are complex indoles representing evolution of an unusual synthesis pathway (Bruneton 833).

The direct extraction of only one ounce of vincristine requires fifteen tons of Periwinkle leaves and costs well over six thousand dollars.

The plant is indigenous to Africa that has been used for several years in East Africa for the treatment of cancer and diabetes. Research on this plant yielded two very important anti-cancer drugs: vinblastine and vincristine. Although the plant is native to Madagascar, it was the US Pharmaceutical firm, Eli Lilly, that commercialized the drugs from germplasm obtained from the Philippines and Jamaica.

Commonly known as the Rosy Periwinkle or Madagascar Periwinkle, the flower has been widely introduced beyond its original island. Velban, the Lilly trade name for Vinblastine, derives from an abundant compound within the plant's leaves. Another cancer drug from the same plant, oncovin, begins as a very scarce alkaloid called Vincristine. Fortunately, due to a process developed in Hungary the more abundant Vinblastine can be transformed in the laboratory to Vincristine. The discovery of both these active substances, now produced commercially by Lilly, can be traced to a large-scale effort of the United States National Cancer Institute in the 1950's to screen plants for anti-cancer properties.

The anti-tumour activity of another Vinca-alkaloid, vinorelbine, is thought to be primarily due to the inhibition of mitotic cell division at metaphase stage through its interaction with micro-tubulin.

The U.S. Food and Drug Administration approved vinorelbine for the treatment of advanced non-small cell lung cancer (NSCLC) in December 1994. This compound has been already launched in more than 70 centuries of the world as a therapeutic agent of non small cell lung cancer and breast cancer.

Before the discovery of vincristine, the survival rate of children with leukemia was about 2 per cent. Now the success rate for this drug is over 80%.

The alkaloids (vinblastine and leucocristine) obtained from *Catharanthus roseus* have found to be effective in curing leukemia (a cancer characterized by excessive number of leucocytes or white blood cells in the body). The drugs manufactured from these alkaloids have received success in curing 50 to 70 per cent cases of some lymphomas. The cases of lymphocytic leukemia, if detected at an early stage, show 99 per cent remission from the drugs having vinblastine derivatives.

Vinblastine, however, reduced the white cell content of the blood. Runaway production of white blood cells is one characteristic of leukemia. Indeed, without vincristine therapy, there would be no cures of childhood leukemia. Also, in combination with other medicines, vinblastine therapy achieves a cure rate of 80 per cent in cancer of the testicle, formerly fatal in most cases. The Periwinkle alkaloids apparently prevent cell division in animals and man.

Vinblastine and vincristine, act like a liquid "sponge" causing the body's blood to absorb extra oxygen. This extra oxygen is especially useful in enriching the brain.

Vincristine is now the drug of choice for the treatment of Childhood leukemia, and it has an annual retail value of over $US 150 million. Vinblastine is used for treating Hodgkin's disease, a serious ailment of unknown origin resulting in enlarged lymph glands, increased numbers of lymphoid tissue cells (hyperplasia) in the spleen, liver and other organs as well as anemia. Before the discovery of vincristine, the survival rate of children with leukemia was about 20 per cent. But a 5 mg ampule of vincristine sulfate injection costs about $US 80 making the drug virtually inaccessible to most leukemia children outside the industrialized world.

Hodgkins disease represents tumours of the lymphatic system. This system is a network of organs, ducts and nodes that interact with the blood's circulatory system to transport a watery clear fluid called lymph throughout the body. The lymphatic system produces lymphocytes, important immune factors and so is a major line of defense against infectious organisms. This system also restores 60% of the fluid that leaks out from blood capillaries back into circulation, and its ducts provide transportation for fats, proteins and other substances that it collects from the body's tissues.

Hodgkin's disease (HD) is the major tumour in a group known as malignant lymphomas. Most often HD starts in B-cell lymphocytes located in lymph nodes in the neck area, although any lymph node may be the site of initial disease.

Hodgkin's disease is more common in males than in females. In the last few years, however, the incidence of HD has increased dramatically in young women.

Age

Hodgkin's disease is the most common malignancy in people of ages 10 to 30 years. The average age for developing Hodgkin's is about 28 years. There are two periods of peak incidence in HD over a life time. The major peak occurs in young people between the ages of 15 and 24 with a lesser peak after age 55. The disease can however, occur at all ages, even in children.

A 49-year old machinist came as a patient to Lilly's Clinical Research Laboratories. For nine months he had suffered from generalized Hodgkin's disease. "When he came to Lilly's he carried a ten – inch tumour in his chest on March 2, 1960, he received his first injection of Lilly's new anti-cancer product, Velban. Within four days his pain had disappeared. In a week he was walking. After 12 days his tumour began to shrink... in four months the tumour was gone" (Ayensee *et al.,* 186). The 'Velban' was the Lilly's trade name for vinblastine.

Chemotherapy and combined modalities for Hodgkin's disease

Chemotherapy uses cell-killing (Cytotoxic) drugs to kill cancer cells throughout the body, and it is therefore referred to as systemic therapy.

Specific agents and Drug combinations used in Hodgkin's Disease

A number of chemotherapy regimens have proven to be effective. Different drugs in each regimen may need to be taken orally or by injection. Toxicity varies depending on the combination. Researchers are investigating the use of chemotherapy regimens used alone, taken in alternate cycles with another regimen, or as hybrids (combining one or two drugs from one regimen with drugs from another).

- ABVD [doxorubicin (Adriamycin), bleomycin, vinblastine, dacarbazine] is now the first choice for most patients who need chemotherapy. It is the most effective and least toxic regimen available to date.
- MOPP [mechlorethamine (Mustargen), vincristine (Oncovin), procarbazine, prednisone] was the standard combination in the past. It carries a 3% risk for leukemia, however, which, if it develops, does so on average five years after treatment.

Other regimens continue to be tested for effectiveness and safety in both early and late stage HD include the following.

- VBM [vinblastine, bleomycin, methotrezate]
- NOVP [mitoxantrone (Novantrone), vincristine (Oncovin), Vinblastine, prednisone], NOVP, may preserve male fertility, and one study reported a five-year survival rate of 95% in stage I and II patents who took it.
- BEACOPP (bleomycin, etoposide, Adriamycin, cyclophosphamide, vincristine, procarbazine, and prednisone) is a very promising regimen, with studies reporting remission rates of over 90% in patients with advanced Hodgkin's. The risk of long-term side effects, such as leukemia, requires longer follow-up.
- A MOPP/ABV hybrid has been investigated. Studies have shown it to be equally effective to the separate drug combinations, but it was highly toxic.
- CHOPE (cyclophosphamide, doxorubicin, vincristine, prednisone, and etoposide). This combination was promising at standard does for advances-stage patients in a 2001 study.
- Stanford V (seven drugs administered over a shorter time but more frequently than MOPP or ABVD). Mechlorethamine (Mustargen), doxorubicin (Adriamycin), vinblastine (Velban), Vincristine (Oncovin), bleomycin (Blenoxane), etoposide (VP-16), prednisone by mouth every other day.

- Three cycles of PABIOE [Adriamycin (doxorubicin), bleomycin, vincristine (Oncovin) and etoposide] that alternate with three cycles of ChlVPP (chlorambucil, vinblastine, procarbazine and prednisone).

Conclusion

Due to reliance on natural sources for the commercial supply, this plant became threatened by over collecting. To conserve it the government of India has placed an embargo on the export of the roots and leaves. In some instances, market demand can threaten the survival of these species in the wild. Yet as with the Periwinkle in China, propagation of medicinal plants in the developing world can reduce the erosion of genetic resources and create business opportunities. There is currently an enormous surge of interest in the use, development and conservation of medicinal plants throughout the world. Several official bodies are involved, including the World Health Organization (WHO) and the OECD. Traditional medicine also has its enthusiasts in the industrial nations, where herbal teas are so popular that some of their plant sources are locally endangered. Low yield of the anti-leukemic alkaloids in the plant combined with their high market price has encouraged intense research.

Acknowledgement

We are grateful to Dr. Mary Chacko, and Dr. Hari N. Iyer, Dept. of Botany, Mar Thoma College, for her encouragement and support during the course of the preparation of this paper. We thank Prof. Thomas Mathew, Head of the Department of Botany, Mar Thoma College, for his co-operation and for providing the required facilities. We are indebted to Mr. Sajan Varghese, Librarian, Mar Thoma College for helping us access the internet.

We wish to place on record the help rendered by Mr. Krishna Kumar, Faculty in the computer section of the college.

References

2001 Nidus information service, Inc.41 East 11th Street, 11th Floor, New York, NY10003ore.mail office@wellconnected.com or on the internet at http:// www. Wellconnected.com.

Shukla, Y.N., Archana Rani and S. Kumar 1997. JMAPS 19 : 430-431.

Taylor, W.I. and N.R. Farnsworth 1975. The Catharanthus alkaloids. Botany, Chemistry, Pharamacology and clinical use, Marcel Dekker. Inc. New York.

BUTTERFLY FAUNA OF AYMANAM GRAMAPANCHAYATH

Jayasooryan K.K., J. Ranjini and A.P. Thomas

School of Environmental Sciences, Mahatma Gandhi University

Abstract

The butterfly fauna of Aymanam Panchayath was studied during 2003-2004. A total of 62 species belonging to the families Papilionidae, Pieriidae, Nymphalidae, Lycaenidae, and Hespiriidae were identified. They showed habitat preferences. The butterflies that showed high occurrence were Common Crow, Lemon Pancy, Common Rose, Common Evening Brown, Grey Pancy etc. The species such as Crimson Rose, Common Jezebel, Tiny Glass Blue, Rice Swift etc showed medium occurrence and Southern Bird wing, which is the largest butterfly in India, Clipper, Striped Albatross, Chocolate Albatross, Giant Orange Tip, Red Pierrot etc showed least occurrence. A decrease in the number of individual species in the agricultural areas was also noticed.

Introduction

The Aymanam Panchayath (Loc: 10^{0}35' N and 74^{0}34' E) lies adjacent to the Vembanad Lake, the largest Ramsar site in Kerala and is traversed by Meenachil River and its branches. Before the construction of Thanneermmukkom barrier (1975) the area was a flood plain of Vembanad Lake. The area has its own importance in the view of backwater tourism. Paddy is the major cultivation and is noted for good riparian vegetation and sacred groves.

The survival and diversity of butterflies in an area depends on availability of food plants and larval host plants, and the plants inturn are benefited by these pollinators. Thus the importance of conservation of butterflies is significant in biodiversity conservation. But the butterfly fauna in Kerala, especially in the rural areas is not studied in detail. This work aims to make a preliminary inventory of butterfly population in this wetland region.

Methodology

The survey was conducted during the monsoon and post-monsoon season in 2003 and 2004. 12 sites were studied. The sampling sites include both wetland and terrestrial ecosystems. Most of the species were identified by visual observation. Species difficult to identify were netted and released after identification. A few species were collected and preserved for further study. The butterflies were identified with the help of standard references (Kunte 2000, Palot *et al.,* 2003, Gunathilakaraj 1998, Larsen 1978 and Gay *et al.* 1992. The species were categorized in to three according to their occurrence such as very common (+++), common (++) and rare (+) (Basistha *et al.*1999).

Result and Discussion

A total of 62 species were identified from the sampling sites (Table 1). They belong to families such as Papilionidae, Pieridae, Nymphalidae, Lycaenidae and Hespiriidae. Lycaenidae family represented largest number of species (Fig. 1). The species very commonly found (Table 1) were Common Crow, Lemon Pancy, Common Rose, Common Evening Brown, Grey Pancy etc. The species having common occurrence were Common Grass Yellow, Blue Mormon, Peacock Pancy, Plain Tiger, Chestnut Bob etc. The rare species noticed were Clipper, Striped Albatross, Chocolate Albatross, Giant Orange Tip, Tamil Yeoman, Red Pierrot, Tawny Coster etc. Southern Bird Wing is the largest butterfly of India (Gunathilakaraj, 1998, Palot, 2003, Kunte, 2000). Tamil Yeoman, Southern Bird Wing and Blue Mormon are endemic to India and Srilanka (Palot, 2003). According Gunathilakaraj (1998) Tamil Yeoman is found above 2500 m in the Western Ghats, but its presence here is noteworthy. Tawny Coster is the only south Indian representative of the subfamily Acraeinae (Palot, 2003, Gunathilakaraj 1998). Some of the butterflies in the present list have been protected by including them in the respective schedules of the

Wild Life Protection Act (WPA) 1972; Crimson Rose in schedule I; Grey Count, Common Baron, Chocolate Albatross and Pea Blue in schedule II; Striped Albatross in schedule IV. Clipper is an endangered as well as rare species (Palot, 2003). The population and diversity of butterflies were high adjacent to the riverine vegetation and in the sacred groves corresponding to the high species diversity of plants. The species like Tamil Yeoman, Grey Count, Blue Mormon, Clipper, Chestnut Bob etc are found in dense greeneries (Palot 2003, Gunathilakaraj, 1998, Kunte, 2000). But the paddy and cultivated field showed less diversity when compared to other habitats. There observed a decreasing trend in the size of individual population of butterflies in the agricultural field. This may be due to the intensive use of pesticides and weedicides (Palot, 2003). The wetlands showed dominance rather than diversity, of a particular species such as Grey Pansy and Peacock Pansy. These species prefer riversides and paddy fields (Palot, 2003). The butterfly population was in accordance with the type of vegetation prevailed in the study area. The water logged areas, dominated by aquatic weeds, were poorer holders of terrestrial plants. This may be the reason why butterfly diversity was poor in these areas. The post-monsoon season met with more diversity than that of the monsoon season. But the Southern Bird Wing was spotted in almost every site during monsoon season; whereas it was rare during post-monsoon season.

The low lands of Kerala are recently under high threat of landscape modification due to population growth, tourism and other infrastructure developmental activities. This in turn results in destruction of wetlands, lowland forests and natural vegetation on which the animal fauna depends. The Aymanam Panchayath also faces the same crisis. Thus urgent conservation measures have to be taken to protect the butterfly diversity of the area.

Table.1 List of Butterflies from Aymanam Gramapanchayath

Common Name	Scientific Name	Family	Occurrence
Blue Mormon	*Papilio polymnestor*	Papilionidae	++
Common Mormon	*Papilio polytes*	,,	+++
Crimson Rose	*Pachliopta hector*	,,	+
Common Rose	*Pachliopta aristolochiae*	,,	+++
Lime Butterfly	*Papilio demoleus*	,,	++
Southern Bird Wing	*Triodes minos*	,,	+
Tailed Jay	*Graphium agamemnon*	,,	++

Common Mime	*Papilio clytia*	,,	+
Common Emigrant	*Catopsilia Pomona*	Pieridae	+++
Mottled Emigrant	*Catopsilia pyranthe*	,,	++
Common Grass Yellow	*Eurema hecabe*	,,	++
Common Jezebel	*Delias eucharis*	,,	+++
Chocolate Albatross	*Appias lyncida*	,,	+
Striped Albatross	*Appias lybithia*	,,	++
Giant Orange Tip	*Hebomoia glaucippe*	,,	+
Psyche	*Leptosia nina*	,,	+++
Caperwhite/Pioneer	*Anaphis aurota*	,,	+
Common Evening Brown	*Melanitis leda*	Nymphalidae	+++
Common Bush Brown	*Mcalesis perseus*	,,	++
Nigger	*Orsotrioena medus*	,,	+
Common Four Ring	*Ypthima huebneri*	,,	++
Common Three Ring	*Ypthima asterope*	,,	++
Common Five Ring	*Ypthima baldus*	,,	++
Tawny Coster	*Acraea violae*	,,	+
Southern Rustic	*Cupha erymanthis*	,,	++
Tamil Yeoman	*Cirrochora thais*	,,	+
Common Sailor	*Neptis hylas*	,,	+++
Common Lascar	*Pantoporia hordonia*	,,	+
Commander	*Modusa procris*	,,	+
Clipper	*Parthinos sylvia*	,,	+
Grey Count	*Tanaecia lepidea*	,,	++
Common Baron	*Euthlia aconthea*	,,	+
Lemon Pancy	*Junonia lemonias*	,,	+++
Peacock Pancy	*Junonia almana*	,,	++
Grey Pancy	*Junonia atlites*	,,	+++
Chocolate Pancy	*Junonia iphita*	,,	+++
Painted Lady	*Cynthia cardui*	,,	+
Glassy Blue Tiger	*Parantica aglea*	,,	+++
Blue Tiger	*Tirumala limniace*	,,	+++
Dark Blue Tiger	*Tirumala septentrionis*	,,	+++
Plain Tiger	*Danaus chrysippus*	,,	++
Striped Tiger	Danaus genutia	,,	++

Common Crow	Euploea core	,,	+++
Common Pierrot	*Castalius rosimon*	Lycaenidae	++
Quaker	*Neopithecops zalmora*	,,	++
Tiny Glass Blue	*Zizula hylax*	,,	++
Pea Blue	*Lampides boeticus*	,,	+
Dark Cerulean	*Jamides bochus*	,,	+
Common Cerulean	*Jamides celeno*	,,	+
Red Pierrot	*Talicada nyseus*	,,	+
Western Centaur Oak Blue	*Arhopala pseudocentaurus*	,,	+
Monkey Puzzle	*Rathinda amor*	,,	+
Indian Sun Beam	*Curetis thetis*	,,	+
Plum Judy	*Abisara echerius*	,,	+
Tricolour Flat	*Pseudocoladenia indrani*	Hesperiidae	+
Dark Palm Dart	*Telicota ancilla*	,,	++
Chestnut Bob	*Iambrix salsala*	,,	++
Coon	*Psolos fuligo*	,,	+
Rice Swift	*Borbo cinnara*	,,	++
Grass Demon	*Udaspus folus*	,,	+
Bush Hopper	*Ampitta dioscorides*	,,	+
Blank Swift	*Baoris kumara*	,,	+

Occurrence: very common: +++, common: ++, rare: +

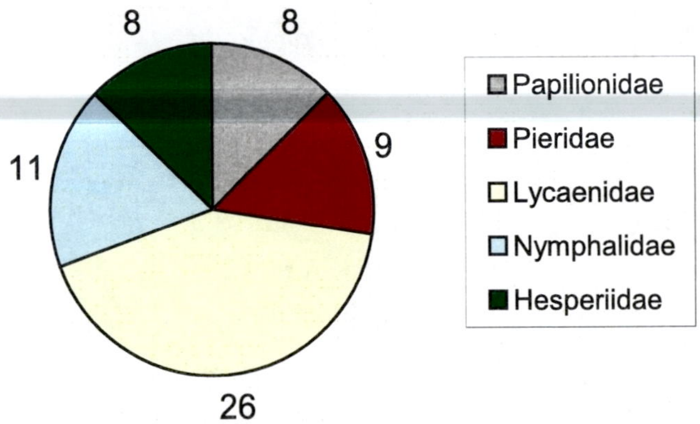

Fig 1. Status of butterfly families in the study areas

References

Basistha, S.K., Ahmed, F and Deka, P. 1999. Butterflies of Orang Wild Life Sanctury, Assam. *J. Zoo's Print.* Vol. XIV (4) pp: 9

Gay, T., Kehimkar, I.D. and Punetha, J.C. 1992. *Common Butterflies of India.* World Wide Fund For Nature – India. Oxford University Press. Bombay.

Gunathilakaraj, K. and Kumar, M.G. 1998. Further addition of Butterflies of Coimbatore City.*J. Zoo's Print.* Vol. XIII (5) pp:16

Kunte, K. 2000. *Butterflies of Peninsular India.* University Press, Hyderabad

Larsen, B.T. 1978: The Butterflies of Nilgiri Mountains of South India (Lepidoptera-Rhopalocera) *J. Bombay Nat. Hist.* Soc.Vol. 84 (3) pp: 26-39

Palot, J., Balakrishnan, V.C. and Kambrath, B. 2003. *Keralathile Chithra-salabhangal.* Mal. Nat. Hist. Soc. Calicut.

A STUDY ON THE ICTHYOFAUNAL DIVERSITY OF AYMANAM PANCHAYAT, IN VEMBANAD WETLAND, KERALA

Thapanjith T., S. Prasanth Narayanan and A.P. Thomas

School of Environmental Sciences, Mahatma Gandhi University, Kottayam.

Abstract

A study on the icthyofauna of the Aymanam Panchayath was carried out and a list was prepared. A total of 37 species of fishes belonging to 18 families and 9 orders were recorded. Order Perciformes showed maximum family diversity .The highest number of species belong to the family Cyprinidae. Of the 34 species, sixteen species are threatened, *Ompok malabaricus* and *Hyporhampus xanthopterus* are critically endangered (CR), *Labeo dussumieri, Horabagrus brachysoma, Tetraodon travancoricus etc.* are endangered (EN), *Puntius sophore, Nandus nandus, Wallago attu* and *Xenentodon cancila* come under lower risk – near threatened (LR-nt) category and *Puntius vittatus, Pristolepis marginatus, Anabas testudineus* etc under vulnerable (VU) category. One exotic species *Poecilia reticulata* was also noted.

Introduction

Forty-one west flowing and three east flowing rivers having a total length of 32,000 km originating from the Western Ghats support Kerala. It is provided with plenty of wetlands including lakes, canals, ponds, paddy fields etc. Also, Kerala being a part of the Western Ghats, one of the 25 hotspots identified around the world, its pisci fauna is very rich. 210 species of freshwater fishes were identified from the Kerala, of this more than 26 species are considered as endemic (Shaji & Nair, 2003). Most of the fishes are characterized by vibrant attractive colours and shades. Some fishes are of ornamental varieties having high potential in the domestic and international market (Radhakrishnan & Kurup, 2002).

The study area is the Aymanam Panchayath. The study was conducted in two season, monsoon and post mansoon periods, of 2003 and 2004. Aymanam Panchayath is located between the two main branches of Meenachil river namely Pennar in the north and Kavanar in the south. There are several canals crisscrossing the Panchayath having Arpookkara Panchayath in the east and Vembanad estuary on the west. Vembanad – Kol wetland has been designated as a Ramsar site in November 2002. Organically rich sedimentary substratum of the inshore regions of the estuary is highly preferred habitat for fish and shrimp breeding (Sreekumar, 2003). Western part of the Panchayath is lying below the MSL, but the eastern part of the land is above MSL.

Methodology

The Ayamanam Panchayath was surveyed for fishes. Twelve sampling sites were fixed in different locations all along the Aymanam Panchayath. Samples were collected by using hook and line method, cast net, gill net and from local fishermen. Unidentified fishes caught during the survey were labelled along with the vernacular names, preserved and brought to the laboratory for further study. The specimens that can be identified at the sites are examined and released. Identification was done with the help of standard reference materials (Day, 1898; Jayaram, 1981 and Daniels, 2002).

Results and discussion

A total of 33 species of freshwater and 4 species of estuarine fishes belonging to 18 families and 9 orders were identified (Table 1 & 2). In the present study family Cyprinidae showed maximum diversity of 29.41% (10 species), and family Bagridae represented by 3 species (8.82%). Exotic species *Poecilia reticulata* were collected from small ditches, which might have been introduced for controlling mosquito larvae. Of the 37 species *Ompok malabaricus* (Goan Catfish) and *Hyporhampus xanthopterus* (Vembanad Halfbeak) are critically endangered (CR), *Labeo dussumieri, Horabagrus brachysoma, Tetradon travancoricus* are endangered (EN) and *Puntius vittatus, Anabas testudineus, Mystus vittatus, Pristolepis marginatus* and

Heteropneutes fossilis come under vulnerable (VU) category. *Parluciosoma daniconius, Puntius sophore, Nandus nandus* and *Xenentodon cancila etc.* come under lower risk – near threatened (LR-nt) category. 50% of the fishes reported from the area are currently considered as non-threatened species (Fig. 1). 11 Western Ghats endemic species including 4 Kerala endemic species were identified. The occurrence of these species in Kerala has already been reported (Ramesh *et al.*, 2003). Of the 37 species of fishes identified, 89.18% (33 species) were having ornamental value (Daniels, 2002; Sekharan & Kurup, 2002 and Nair & Shaji, 2003).

All, except *Aplocheilus blochii, A. lineatus, Poecilia reticulata, Chelenodon fulviatilis* and *Tetraodon travancoricus* are used as food fishes. Among the fishes used as food, *Etroplus suratensis* is high priced. As per the report of the locals *Tetraodon travancoricus* and *Chelenodon fulviatilis* are not even used as poultry feed. It is poisonous due to the presence of bacteria living inside them (Mathai, Pers. Comm. 2003). However, detailed study is needed to confirm the poisonous identity of the species *Tetradon travancoricus* and *Chelenodon fulviatilis*. In the present study critically endangered (CR) Channa micropeltes and estuarine species such as *Megalops cyprinids, Scatophagus argus, Lutjanus johnii, Lates calcarifer* could not been collected. According to the local fishermen and people the above-mentioned species are also found in the study area.

According to the locals, Thannermukkom salt water barrage, the unscientific fishing (poisoning) and the increased pollution of the Vembanad lake due to tourist boats and hotels caused the decline of the fish diversity of the area, as the wetland waters here are closely connected with the Vembanad lake. The fish disease, Epizootic Ulcerative Syndrome has constantly affected the fishes of this region. Once the presence of mangrove plants and swamps favoured the egg laying activity of fishes, but because of the destruction of mangrove vegetation and pollution load, now the fish fauna of this wetland is under threat. Further studies are needed to make the complete checklist of the fish species of the Aymanam Panchayath.

Table 1. Systematic List of Species from Eastern part of Aymanam Panchayath

Order : Cypriniformes

Family : Cyprinidae

1. *Danio malabaricus* (Jerdon) — Giant Danio
2. *Rasbora daniconius* (Hamilton) — Blackline Rasbora

3. Amblypharyngodon melittinus (Valenciennes)	Attentive Carplet
4. *Puntius amphibious* (Valenciennes)	Scarletbanded Barb
5. *Puntius filamentosus* (Valenciennes)	Blackspot Barb
6. *Puntius sarana* subnastus (Valenciennes)	Peninsular Olive Barb
7. *Puntius sophore* (Hamilton)	Softfin Barb
8. *Puntius vittatus* (Day)	Koolie Barb
9. *Labeo dussumieri* (Valenciennes)	Small scaled Labeo
10. *Garra mullya* (Sykes)	Brown Algae Eater

Order : Siluriformes

Family : Bagridae

11. *Mystus gulio* (Hamilton)	
12. *Mystus vittatus* (Bloch)	Striped-dwarf Catfish
13. *Horabagrus brachysoma* (Day)	Golden Catfish

Family : Siluridae

14. *Ompok malabaricus* (Valenciennes)	Goan Catfish
15. *Wallago attu* (Schneider)	Boal

Family : Heteropneustidae

16. *Heteropenustes fossilis* (Bloch)	Stinging Catfish

Order : Mugiliforms

Family : Mugilidae

17. *Mugil cephalus* (Linneaus)	Mullet

Order : Beloniformes

Family : Belonidae

18. *Xenentodon cancila* (Hamilton)	Freshwater Garfish

Family : Hemirhamphidae

19. *Hyporhamphus xanthopterus* (Valenciennes)	Vembanad Halfbeak

Order : Cyprinidontiformes

Family : Aplocheilidae

20. *Aplocheilus lineatus* (Valenciennes)	Tiger Panchax

21. *Aplocheilus blockii* (Arnold) Little Panchax

Family : Poecilidae

22. *Poecilia reticulata* (Peters) Guppy

Oreder : Synbranchiformes

Family : Mastacembelidae

23. *Mastacembelus armatus* (Lacepede) Tyre-trackspiny Eel

Order : Perciformes

Family : Nandidae

24. *Nandus nandus* (Hamilton) Mottled Leaffish
25. *Pristolepis marginatus* (Jerdon) Malabar Catopra

Family : Cichlidae

26. *Etroplus suratensis* (Bloch) Banded Pearl Spot
27. *Etroplus maculates* (Bloch) Orange Chromide

Family : Gobiidae

28. *Glossogobius giuris* (Hamilton) Tank goby
29. *Awaous* spp.

Family : Anabantidae

30. *Anabas testudineus* (Bloch) Indian Climbing Perch

Family : Belontidae

31. *Macropodus cupanus* (Valenciennes) Indian Paradisefish
32. *M.cupanus dayi* (Day) Day's Paradisefish

Family : Channidae

33. *Channa striatus* (Bloch) Banded Snakehead
34. *Channa marulius* (Hamilton & Buchnan) Giant Snakehead

Order : Pleuriectiformes

Family : Soleidae

35. *Synaptura orientalis* (Bloch& Schneider) Estuarine Sole

Order : Tetradontiformes

Family : Tertradontidae

36. *Tetradon travancoricus* (Hora & Nair) Little Pufferfish

37. *Chelenodon fulviatilis* (Hamilton & Buchnan) Green Pufferfish

Fig 1. Status of Fishes in the Study Area

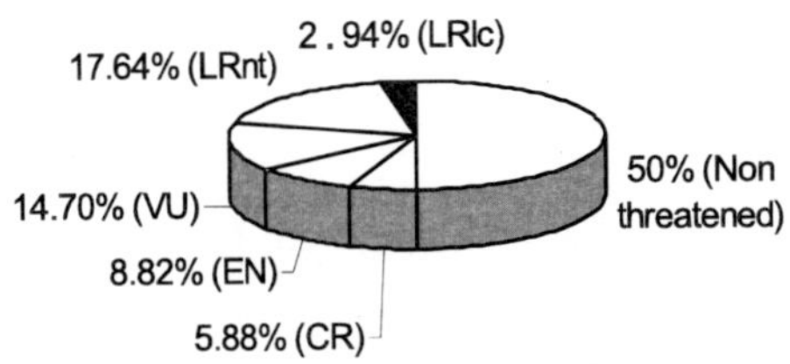

Table 2. List of Fishes and its Categories

Sl. No.	Scientific Name	Common Name	Vernacular Name	Endemism	IUCN Status	Economic Importance
1	*Danio malabaricus*	Giant Danio	Vayampu, Paral	EN-WG		Eatable, Aquarium species
2	*Rasbora daniconius*	Blackline Rasbora	Thuppal-ukothi, Thuppal-umathi		LRnt	Eatable, Aquarium species
3	*Puntius amphibious*	Scarletbanded Barb	Urulanparal	EN-SI		Eatable, Aquarium species
4	*Puntius filamentosus*	Blackspot Barb	Kotichip-paral, Poova-lipparal	EN-WG		Eatable, Aquarium species
5	*Puntius sarana*	Peninsular Olive Barb	Kuruva			Eatable, Aquarium species
6	*Puntius sophore*	Softfin Barb	Paral		LRnt	Eatable, Aquarium species
7	*Puntius vittatus*	Kooli Barb	Vattakkali, Kaippa		VU	Poultry feed, Aquarium species & eatable
8	*Labeo dussumieri*	Kerala Labeo	Pullan		EN	Eatable, Cultivable
9	*Garra mullya*	Brown Algae Eater	Kallemutti, Kallelokkan	EN-I		Eatable, Aquarium species
10	*Amblypharyngodon melittinus*	Attentive Carplet	Vayampu			Eatable, Aquarium species, Poultry feed
11	*Horabagrus brachysoma*	Yellow Catfish	Manjakkori	EN-K	EN	Eatable, Cultivable and Aquarium species

12	*Mystus gulio*	Longwhiskered Catfish	Vellakkori			Aquarium species, Eatable
13	*Mystus vittatus*	Striped Dwarf Catfish	Chillan		VU	Aquarium species, Eatable
14	*Ompok malabaricus*	Goan Catfish	Thlappan	EN-WG	CR	Eatable
15	*Wallago attu*	Boal	Valah		LRnt	Aquarium species, Eatable
16	*Heteropneustes fossilis*	Stinging Catfish	Kaari		VU	Aquarium species, Eatable
17	*Mugil cephalus*	Mullet	Kanampu			Eatable and Cultivable
18	*Xenentodon cancila*	Freshwater Garfish	Kola, Kolan		LRnt	Eatable, Aquarium species
19	*Hyporhampus xanthoptrus*	Vembanad Halfbeak	Morasu	EN-K	CR	Eatable, Aquarium species
20	*Aplocheilus lineatus*	Tiger Panchax	Poonjan	EN-WG		Poultry feed, Aquarium species
21	*Aplocheilus blockii*	Little Panchax	Poonjan	EN-WG		Poultry feed, Aquarium species
22	*Poecilia reticulata*	Guppy		EXO		Aquarium species
23	*Mastacembelus armatus*	Tyre- trackspiny Eel	Aarakan			Aquarium species, Eatable
24	*Nandus nandus*	Mottled Leaffish	Muthuvala		LRnt	Eatable, Aquarium species
25	*Pristolepis marginataus*	Malabar Catopra	Pannaka-rimeen	EN-K	VU	Aquarium species, Eatable
26	*Etroplus suratensis*	Banded Pearl Spot	Karimeen, Kariya-mplachi			Eatable, Aquarium species
27	*Etroplus maculates*	Orange Chromide	Pallathi			Eatable, Aquarium species
28	*Glossogobius giuris*	Tank Goby	Poolan			Aquarium species, Eatable
29	*Awaous spp.*		Pukalap-poolan			Aquarium species, Eatable
30	*Anabas testudineus*	Climbing Perch	Kallada		VU	Eatable, Aquarium species
31	*Macropodus cupanus*	Indian Paradisefish	Karimkana			Aquarium species, Poultry feed
32	*M. cupanus dayi*	Day's Paradisesfish	Karimkanna			Aquarium species, Poultry feed
33	*Channa striatus*	Banded Snakehead	Varal		LRlc	Eatable, Aquarium species
34	*Channa marulius*	Giant Snakehead	Cherumeen		LRnt	Eatable, Aquarium species

35	*Synaptura orientalis*	Estuarine Sole	Kayal Nanku			Eatable
36	*Tetradon travacoricus*	Little Pufferfish	Vaalmakri	EN-K	EN	Aquarium species, Eatable
37	*Chelenodon fulvialilis*	Green Pufferfish	Oothive-erppan		----	Aquarium species

(Endemism and Status are based on Fresh Water Fishes of India, Report Summary 1998 and A Brief outline of Biodiversity Conservation and Action Plan for Kerala, Ramesh *et al.*, 2003)

EN-WG - Endemc to Western Ghats EN-SI - Endemic to South India
LRnt - Lower risk near threatened EN- K – Endemic to Kerala
EN - Endangered LRlc - Lower risk least concern
EN-I - Enndemic to India VU - Vulnerable EXO - Exotic

References

Bevan, R. 1982. *Handbook of the Freshwater Fishes of India.* Narendra Publishing House. New Delhi.

Daniels, R.J.R. 2002. *Fresh Water Fishes of Peninsular India.* Universities Press Pvt. Ltd.

Day, F. 1989. The Fauna of British India Including Ceylon and Burma Fishes. Vol. I and II. Today and Tomorrow's Printers. New Delhi. Second Reprint.

Fresh Water Fishes of India, Report Summary 1998. 1999. *Zoo's Print Jour* 14(3).

Jayaram, K.C. 1999. *The freshwater Fishes of the Indian Region.* Narendra Publishing House. Delhi.

Jinghran, V.G. 1975. *Fish and Fisheries of India.* Hindustan Publishing Co oporation. Delhi.

Menon, A.G.K. 1997. Rare and Endangered Fishes of Malabar, India. *Zoo's Print Jour.*

Mercy, T.V.A., R. Thomas, and E. Jacob 2002. Potential Ornamental Fishes of the Western Ghats. In: *Riverine and Reservoir Fisheries of India.* Society of Fisheries Technologists. Cochin.

Padmakumar, K.G. 2003. Conservation and Management of Vembanad Wetlands, Kerala. In: Vembanad Water Bird Count (Sreekumar, Ed.). Dept. of Forest and Wildlife, Govt. of Kerala.

Radhakishnan, K.V and B.M. Kurup 2002. Habitat Preference and Biodiversity of Potential Freshwater Ornamental Fishes of Kerala. In: *Riverine and Reservoir Fisheries of India.* Society of Fisheries Technologists. Cochin.

Ramesh, B.R., P.V. Karunakaran, M. Balasubramanian, L.D. Seen and O.P. Kaler 2003. *A Brief Outline of Biodiversity Conservation and Action Plan for Kerala.* Kerala Forest and Wild Life Dept. Kottayam.

Sekharan, M., P.K. Pramod and A. Ramachandran 2002. Role of Chalakkudy Riverine Resources in the Ornamental Fish Marketing Sector of Kerala.

In: *Riverine and Reservoir Fisheries of India.* Society of Fisheries Technologists. Cochin

Sreekumar, B. (Ed.) 2002. *Vembanad Water Bird Count*. Dept. of Forest and Wildlife, Govt. of Kerala. Kottayam.

A COMPARATIVE STUDY ON THE NUMBER OF WETLAND BIRDS IN THE KTDC TOURIST COMPLEX, KUMARAKOM

Sreekumar B.[1] and S. rasanth Narayanan[2]

1. Ornithologist, Kottayam Nature Society. Kottayam.
2. School of Environmental Sciences, Mahatma Gandhi University, Kottayam

Abstract

Kerala Tourism Development Corporation (KTDC) Tourist Complex of Kumarakom of the Kottayam district has a few relict populations of mangrove species. So far 17 mangrove species were reported from Kerala (Unni, 2002). In which, KTDC Tourist Complex of Kumarakom have 6 true mangrove species and 17 marshy associate species (Ravi, 2003). This supports a good number of birds and it is one of the largest heronries in Kerala. Out of 483 species of all birds over Kerala, 39.13% (189 species) of bird species were reported from Kumarakom and Vembanad region (Sreekumar, 2003). A total of 15 species coming under 6 families and 2 orders were identified. Order Ciconiformes showed maximum family diversity and family Aredeiadae showed maximum species diversity. Globally threatened species Darter *Anhinga melanogaster* is nesting and breeding here and found all through these four years. Reports of 'Vembanad water bird count' of four years were compared.

Introduction

'Gods Own Country'- Kerala comprises a narrow southernmost strip of land, its greatest length from north to south is about 545 km and greatest width is 120 km (Ali 1984). Kerala is blessed with high biological diversity. Lowlands of Kerala is blessed with beautiful estuaries, lakes and rivers. Once the coastal and estuarine areas of Kerala were covered by 700 km^2 of mangroves. But today mangrove areas of Kerala shrinked to 50 km^2 (Nayar *et al.,* 2002). Available data on the district wise distribution of mangroves of Kerala shows that more area under mangroves still exists in the northern districts, particularly Kozhikkode and Kannur. Kerala Tourism Development Corporation (KTDC) Tourist Complex of Kumarakom of the Kottayam district possesses relict populations of mangroves species.109 species of mangroves and mangrove associates were identified from India (Deshmukh & Rao 2003). So far 17 mangrove species were reported from Kerala. In which, KTDC Tourist Complex of Kumarakom has 6 mangrove species and 17 marshy associate species (Ravi, 2003).

483 species of birds were reported from Kerala (Neelakantan 1996). Out of 483 species all over Kerala, 39.13% (189 species) of bird species were reported from Kumarakom and Vembanad region (Sreekumar 2003). Wetlands are important bird habitats, and birds use them for breeding, nesting, and rearing young (Stewart, 2001). Water birds are an important component of wetlands and occupy several trophic levels in the food web; thus help maintain the heterogeneity of the ecosystem. Wetland birds are broadly defined as 'birds ecologically dependent on wetlands' and include recognised groups popularly known as wildfowl, waterfowl, shorebirds and waders (Jayson, 2002).

Study area

KTDC Tourist Complex of Kumarakom lies between E 76°25' and 76°26' longitudes and N 9°37' and 9°38' latitudes and lies below MSL. It is a piece of land that measures 112 acres, 14 km west of Kottayam town. It has Kumarakom – Vechoor road on the east, Kavanar River on the north, the farm of the Regional Agricultural Research Station of Kerala Agricultural University in the south and freshwater dominating southern part of the Vembanad estuary on the west. Vembanad–Kole wetland has been designated as a Ramsar site in November 2002. The estuarine zone near the tourist complex containing organically rich sedimentary substratum of the inshore region make it a highly preferred habitat for the breeding shrimp. The primary vegetation shows mangroves such as *Avicennia officinalis, Bruguiera gymnorrhiza, Rhizophora mucronata, Sonneratia caseolaris,* marshy mangrove associates, non-mangroves and hydrophytes (Ravi, 2002). Some of the native vegetation was converted to plantations of coconut and rubber.

Background of the study area

Previously this land was known as Baker Estate. During 1970's and 1980's the Baker Estate was the only place in Kerala where the Night Herons *Ncyticorax nycticorax* bred. Neelakantan (1996) recorded 3000 to 4000 Night Herons *Ncyticorax nycticorax* breeding in the marshy area of the northwest part of the estate. The near threatened bird, Darter *Anhinga melanogaster* (Grewal *et al.,* 2002), other two species of herons, three species of egrets and two species of cormorants are also reported (Sreekumar, 2001). According to Neelakantan (1996) this heronary was the biggest of its kind in Kerala. Now the land is in the hands of KTDC. After the take over by the KTDC there was a proposal to clear off the mangrove belt along the backwater and construct cottages, beer parlour etc. Due to the sustained protest from the public, Kerala State Science and Technology Committee and Prof. V.K. Damodaran visited the place and recommend detailed study on the flora and fauna of the Estate. The Centre for Earth Science Studies (CESS) entrusted with the study, submitted the report and recommended the Government of Kerala to protect the area as such. Following the study the Wildlife Advisory Board of Kerala, in 1989 recommended the Government to declare the Baker Estate as a Bird Sanctuary. Against the scientific advice, 13 hectares of the land was given to a venture company. The company cleared the mangrove and converted it into a lawn and constructed a tourist jetty. With this, a species of mangrove called *Kandelia candel* was totally wiped out (Ramachandran & Mohanan 1990). During the 1990's they cleared bamboo clumps, trees, half of the mangroves and installed high voltage electric lights on the trees. These activities directly affected the breeding birds and as a result one third of the birds left the place (Sugathan pers.com.).

Objectives and methods

The main objective of the study is to find out the change in the pattern of avian distribution in the study area. The midwinter water bird count was done regularly in the KTDC Complex (Sreekumar 1987 to 2000, Anon 1993 and Sreekumar 2001, 2002, 2003, 2004). A comparison of occurrence of wetland avifauna and number deviation of the KTDC Tourist Complex from 2001 – 2004 were done with respect to the "Vembanad Water Bird Count". Kottayam Nature Society in association with the Forest and Wildlife Department conducted water bird count in the KTDC Tourist Complex during the third week of January every year, 2001 – 2004. The counts were done in the early morning, 6AM to 10AM. Direct observations were done through the binoculars and spotting scopes. For observation, team walked along the bunds or foot tracks and counted the wetland birds. The directions of the Wetland International for the 'Mid-Winter Water Fowl Count' were strictly adhered for the count.

Result

A total of 72 species of birds were reported, including fifteen species of wetland birds during these four-years of study, excluding one unidentified crake species. In 2001 the number of birds counted was 198, in 2002 it was reduced to 91, 2003 it showed a slight increase of 96 and in 2004 it increased to 158 (Fig. 1). Globally threatened species Darter *Anhinga melanogaster* and White Ibis *Threskiornis melnocephalus* (Grewal *et al.,* 2002 and Pande, 2003) were also spotted and counted during the 2001-2004 (Table 1). Common wetland species found in the tourist complex during the study period were Darter *Anhinga melanogaster,* Blackcrowned Night Heron *Nycticorax nycticorax,* Purple Heron *Ardea purpurea* and Indian Pond Heron *Ardeola grayii.* In 2001, 65 individuals of globally threatened Darter were counted but in the following three years its population was reduced. Compared to 2001 report Darter population declined to 84.62%, 64.62% and 35.39% respectively during 2002, 2003 and 2004. The globally threatened species, White Ibis was spotted during the study, but the count gradually decreased as the year progressed.

The number of Blackcrowned Night Herons *Nycticorax nycticorax* inside the tourist complex also showed reduced number. Black Bitterns *Dupetor flavicollis* were spotted during 2002 and 2003 but in 2004 count there was no Black Bittern at all. One transcontinental migrant wetland species spotted in the counting area is Glossy Ibis *Plegadis falcinellus.* Indian Shag *Phalacrocorax fuscicollis,* Little Cormorant *Phalacrocorax niger,* Glossy Ibis and Eurasian Marsh Harrier *Circus aeruginosus* are the four species that showed their maximum individual number in 2004. Another wetland dependent transcontinental migratory raptor counted inside the tourist complex is Eurasian Marsh Harrier. The present study was carried out with the aim of listing the status of birds in the KTDC complex for a period of 2001 –2004.

Kumarakom Tourist Complex is the biggest heronry ever found in Kerala. Eventhough bird count is less during the winter; it will be activated with breeding wetland birds during monsoon. The result shows a great reduction in the status of the wetland birds in the study area. Two globally threatened species (Darter and White Ibis) were spotted. The Darter count in 2001 was 0.65% of the world darter population. Darter's global population is estimated as 10,000. Hence, any place that has 1% of the world population of this species will be declared as an Important Bird Area (Rahmani, 2002). The number of Blackcrowned Night Herons also showed reduction in number. Prof.K.K. Neelakantan, during his first visit to the tourist complex in March 1981, has recorded about 3000 Night Herons in the area. During 1970's and 1980's the Baker Estate was the only place in Kerala where the Night Herons *Ncyticorax nycticorax* bred. Cattle Egrett *Bubulcus ibis,* was located only in low numbers, which was comparatively much lower than the study conducted in 1993 (Anon.). Kumarakom tourist complex possess a relict wealth of mangroves of six species. The mangrove flora of the area is gradually dying due the lack of regeneration (Mathew *pers.com*). According to the locals, Thannermukkom salt-water barrage, tourist boats and resorts of the area has enhanced the pollution of the Vembanad Lake.

Recommendations

1. As a priority, ecological reasons should be identified to determine drastic reduction in the population of many groups of Wetland birds. Further studies would be required for this purpose.
2. A detailed study on the movements of the birds of this area should be conducted during various seasons to determine the special and temporal pattern of the bird migration.
3. The KTDC Tourist Complex - Kumarakom has to be declared as a bird sanctuary or a protected area.
4. A new model of protected area, involving the Forest Department, local Panchayath.

Table 1. Bird count 2001 - 2004

Sl. No.	Species	2001	2002	2003	2004
1	Indian Shag	2	2	O	36
2	Little Cormorant	6	7	2	16
3	Darter	65	10	23	42
4	Black Bittern	0	2	2	0
5	Blackcrowned Night Heron	21	19	23	8
6	Indian Pond Heron	12	12	8	8
7	Cattle Egret	0	4	1	4

8	Little Egret	0	6	0	4
9	Intermediate Egret	0	3	0	3
10	Purple Heron	88	6	28	10
11	White Ibis	0	9	4	8
12	Glossy Ibis	0	1	0	4
13	Whitebrested Waterhen	3	6	4	4
14	Brahminy Kite	1	3	0	4
15	Eurasian Marsh Harrier	0	1	0	8
	Total	198	92	96	158

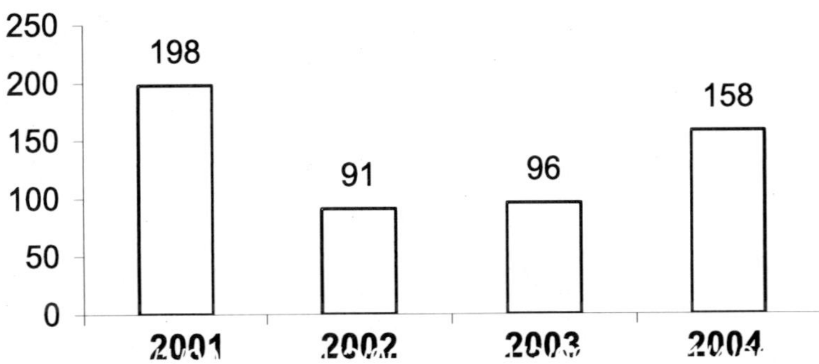

Fig 1. Bird count in the study period.

Checklist of the birds of KTDC Tourist Complex - Kumarakom

Order: Anseriformes
Family: Dendrocygnidae

1. Lesser Whistling Teal *Dendrocygna javanica* (Horsfield)

Family: Anatidae

2. Cotton Teal *Nettapus coromandelianus* (Gmelin)*

Order: Piciformes
Family: Picidae

3. Lesser Goldenbacked Woodpecker *Dinopium benghalense* (Whistler & Kinnear)

Family: Megalaimidae

4. Smallgreen Barbet *Megalaima viridis* (Boddaert)

Order: Coraciformes

Family: Halcyonidae

5. Storkbilled Kingfisher *Halcyon capensis* (Linnaeus)
6. Whitebreasted Kingfisher *Halcyon smyrnensis* (Boddaert)

Family: Alcedinidae

7. Three toed Kingfihser *Ceyx erithacus* (Linneaus)
8. Common Kingfisher *Alcedo atthis* (Kleinschmidt)

Family: Cerylidae

9. Lesser Pied Kingfisher *Ceryle rudis* (Whistler & Kinnear)*
10. Bluetailed Bee-eater *Merops philippinus* (Linnaeus)*

Order: Cuculiformes

Family: Cuculidae

11. Asain Koel *Eudynamis scolopacea* (Linnaeus)

Family: Centropopidae

12. Crow Pheasant *Centropus sinensis* (Streseman)

Order: Psittaciformes

Family: Psittacidae

13. Roseringed Parakeet *Psittacula krameri* (Bechstein)*
14. Plumheaded Parakeet *Psittacula cyanocephala* (Forster)*
15. Indian Lorikeet *Loriculus vernalis* (Sparrman)*

Order: Apodiformes

Family: Apodidae

16. Cochinchina Spinetail Swift *Hirundapus cochinchinensis* (Oustalet)*
17. Alpine Swift *Tacymarptis melba* (Koelz)*
18. House Swift *Apus affinis* (J.E.Gray)*

Order: Strigiformes

Family: Tytonidae

19. Barn Owl *Tyto alba* (Hartert)

Family: Strigidae

20. Jungle Owlet	*Glaucidium radiatum* (Blyth)
21. Brown Hawk Owl	*Ninox scutulata* (Temminck)
22. Collared Scops Owl	*Ottus bakkamoena* (Pennant)

Order: Columbiformes

Family: Columbidae

23. Blue Rock Pigeon	*Columba livia* (Strickland)*
24. Jerdon's Imperial Pigeon	*Ducula badia* (Jerdon)*

Order: Gruiformes

Family: Rallidae

25. Whitebreasted Waterhen	*Amaurornis phoenicurus* (Pennant)

Order: Ciconiformes

Family: Charadridae

26. Blackwinged Stilt	*Himantopus himantopus* (Linneaus)*
27. Redwattled Lapwing	*Vanellus indicus* (Boddaert)

Family: Laridae

28. Whiskered Tern	*Childonias hybridus* (Stephens)*

Family: Accipitridae

29. Osprey	*Pandion haliaetus* (Linneaus)*
30. Brahminy Kite	*Haliastur indus* (Boddaert)
31. Pariah Kite	*Milvus migrans* (Sykes)*
32. Eurasian Marsh Harrier	*Circus aeruginosus* (Linnaeus)
33. Shikra	*Accipiter badius* (Gmelin)
34. Steppe Eagle	*Aquila nipalensis* (Hodgson)*

Family: Anhingidae

35.Oriental Darter	*Anhinga melanogaster* (Pennant)

Family: Phalacrocoracidae

36. Little Cormorant	*Phalacrocorax niger* (Vieillot)
37. Indian Shag	*Phalacrocorax. fuscicollis* (Stephens)

38. Large Cormorant *Phalacrocorax carbo* (Shaw)

Family: Ardeidae

39. Little Egret *Egretta grazetta* (Linnaeus)
40. Purple Heron *Ardea purpurea* (Mayen)
41. Intermediate Egret *Mesophoyx intermedia* (Wagler)
42. Large Egret *Casmerodius albus* (J.E.Gary)
43. Cattle Egret *Bubulcus ibis* (Boddaert)
44. Indian Pond Heron *Ardeola grayii* (Sykes)
45. Night Heron *Nycticorax nycticorax* (Linneaus)
46. Black Bittern *Dupetor flavicollis* (Latham)
47. Chestnut Bittern *Ixobrychus cinnmmomeus* (Gmelin)

Family: Ciconidae

48. Openbill Stork *Anastomus oscitans* (Boddaert)

Family: Threskiornithidae

49. White Ibis *Threskiornis melanocephalus* (Latham)
50. Glossy Ibis *Plegadis falcinellus* (Linnaeus)*

Order: Passeriformes

Family: Corvidae

51. Indian Treepie *Dendrocitta vagabunda* (Whistler & Kinnear)
52. House Crow *Corvus splendens* (Madarasz)
53. Jungle Crow *Corvus macrorhynchos* (Sykes)
54. Ashy Shrike Swallow *Artamus fuscus* (Vieillot)*
55. Black Drongo *Dicrurus macrocercus* (Vieillot)
56. Greater Rackettailed Drongo *Dicrurus paradiseus* (Linneaus)
57. Black Headed Oriole *Oriolus xaanthornus* (Linnaeus)
58. Eurasian Golden Oriole *Oriolus oriolus* (Sykes)
59. Asian Paradise Flycather *Terpsiphone paradisi* (Linneaus)

Family: Muscicapidae

60. Oriental Magpie Robin *Copsychus saularis* (Sclater)

Family: Sturnidae

61. Common Myna	*Accridotheres tristis* (Linneaus)*
62. Jungle Myna	*Accridotheres fuscus* (Sykes)*
63. Grey Headed Myna	*Sturnus malabaricus malabaricus* (Gmelin)

Family: Hirundinidae

64. Common Swallow	*Hirundo rustica* (Scopoli)*

Family: Pycnonotidae

65. Redvented Bulbul	*Pycnonotus cafer* (Linneaus)

Family: Sylvidae

66. Common Tailor Bird	*Orthotomus sutorius* (Latham)
67. Blyth's Reed Warbler	*Accrocephalus dumetorum* (Blyth)
68. Greenish Leaf Warbler	*Phylloscopus trochiloides* (Blyth)

Family: Nectarinidae

69. Tickle's Flowerpecker	*Dicaem erythrorhynchos* (Latham)
70. Purplerumped Sunbird	*Nectarinia zeylanica* (Hermann)
71. Loten's Sunbird	*Nectarinia lotenia* (Whistler)

Family: Passeridae

72. *Forest Wagtail*	*Dendronanthus indicus* (Gmelin)

* observed only in flight

References

Ali, S. 1984. *Birds of Kerala*. Oxford University Press. New Delhi.

Anon 1993. Birds of Vembanad Lake: a survey report. Nature Education Society, Vembanad Nature Club and Kerala Forest and Wildlife Department.

Deshmukh, S and Rao, T.A 2003. Mangrove ecosystems of India: status and policy. *In Sustainable management of wetlands biodiversity and beyond.* Sage publishers India Pvt. Ltd.

Grewal, B, Harvey, B. and Pfister, O. 2002. *A photographic guide to birds of India and the Indian subcontinent*. Periplus Edition (HK) Ltd. Singapore.

Jayson, E.A. 2002. Avifauna in the wetlands of Kerala. In: *Wetland conservation and management in Kerala.* State Committee on Science, Technology and Environment, Thiruvanathapuram.

Neelakantan, K.K. 1996. *Keralathile pakshikal.* Kerala Sahithya Academy. Trichur.

Pande, S. 2003 *Birds of Western Ghats, Kokan and Malabar (including birds of Goa).* Bombay Natural History Society and Oxford University Press.

Rahmani, A. (Ed.) 2002. Asian water bird census: final report, India. Bombay Natural History Society. Bombay.

Ramachandran, K.K. and Mohanan 1990. The mangrove ecosystems of Kerala: final report. CESS. Trivandrum.

Ravi, 2002. Flora of Kumarakom tourist complex. *In:* Vembanad water bird count 2002. Kerala Forest and Wildlife Department. Kottayam.

Sreekumar, B. (Ed.) 2001. Vembanad water bird count 2001. Kerala Forest and Wildlife Department. Kottayam.

Sreekumar, B. (Ed.) 2002. Vembanad water bird count 2002. Kerala Forest and Wildlife Department. Kottayam.

Sreekumar, B. (Ed.) 2003. Vembanad water bird count 2003. Kerala Forest and Wildlife Department. Kottayam.

Sreekumar, B. (Ed.) (In press). Vembanad water bird count 2004. Kerala Forest and Wildlife Department. Kottayam.

Stewart, R E. 2001. Technical Aspects of Wetlands - Wetlands as Bird Habitat. National Water Summary on Wetland Resources. United States Geological Survey.

PHYTOTOXICITY OF MERCURY ON SEED GERMINATION IN *VIGNA RADIATA* (L.) WILCZEK

P. Nisha and S. Sreedevi**

* Post graduate student, Department of Botany, S.N. College, Kollam - 691 001, India

** Senior Lecturer, Department of Botany, S.N. College, Kollam - 691 001, India

Abstract

The present investigation evaluates the toxic effects of mercuric chloride on seed germination in *Vigna radiata*. Seeds of *Vigna radiata* were subjected to varying concentrations of mercuric chloride (10 μm, 50 μm, 100 μm, 250 μm, 500 μm, 750 μm, 0.001M, and 0.01M), with distilled water soaked seeds forming the control. Germination percentage remained high and inhibitory effects were noted only at concentrations above 500 μm. The metal at all concentrations reduced seedling vigor and the effect was more pronounced in the roots as evidenced by inhibition of root growth, and reduced root tolerance index. The % phytotoxicity increased with increase in concentrations of mercuric chloride. Ascorbic acid was found to ameliorate the inhibitory effects of mercury on seed germination. Estimation of the total protein content at all stages of analysis (24, 48, 72 and 96 hours after germination) showed increments over the control. Accumulation of proline, an indicator of stress occurred at higher levels of mercury. Determination of chlorophyll content, leaf area, and total seedling length on the 7th

day indicated reduction in total chlorophyll content, leaf area, and seedling length in treated plants in comparison with the control. Root morphology was found to be adversely affected by mercury as shown by inhibition in growth, thickening and blackening of roots of treated plants.

Introduction

Mercury a divalent group II B metal is a potential biocide to aquatic and terrestrial fauna and flora even at fairly low concentrations. Through industrial, municipal, domestic and agricultural effluents, mercury is introduced into soil and water bodies and transferred through the food chain. The possible injurious effect of exposure of metals to man animals and plants has generated global concern. Information regarding the effects of mercury on crop plants is limited. Therefore this investigation was undertaken to evaluate the effects of $HgCl_2$ on seed germination and growth in green gram *Vigna radiata.*

Materials and methods

Green gram (*Vigna radiata* L. Wilczek) seeds were spread in Petri dishes lined with filter paper soaked in measured quantities of varying levels of mercuric chloride in distilled water soaked seeds formed the control. The breaking of the testa was considered as the criterion for calculation of germination % Sampling was done at intervals of 24, 48, 72, and 96 hours of germination Protein content was estimated using Folin ciocalteau reagent based on Lowry *et al.* (1951). Growth of seedling was assessed in terms of seedling vigor according to Woodstock (1969) and Abdul - Baki and Anderson (1973), Tolerance index was based on Turnur and Marshal (1972) and % Phytotoxicity as per Chou *et al.* (1972). Proline accumulation was determined following the method of Bates *et al.* (1973), measurements of leaf area, estimation of total chlorophyll content, chlorophyll a chlorophyll b was done as per Arnon (1949) on the 7th day

The concentrations of mercuric chloride which were inhibitory were selected to analyze the possible ameliorating effects of antioxidant ascorbic acid on germination of treated seeds.

Results and discussion

The germination % remained high even at 500μm $HgCl_2$, but gradual inhibitory effects were noted at higher levels with 0.01M $HgCl_2$ showing only 50% germination (Fig. 1). The deleterious and adverse effects of mercury was evident in the form of reduction in root and shoot length of all treated samples in comparison with the control. In no case was shoot growth inhibited to the same extent as root growth. High concentrations resulted in shortening and malformation of the roots, which with passage of time became brown, brittle, short and stubby. Less severe effects of the metal on the shoot was most likely due to its weak mobility and hence lower mercury contents of the above ground organs, moreover the roots are the primary sites of metal absorption and accumulation. Significant reduction in seedling vigor, root tolerance index and increase in % phytotoxicity was noted in all the treated samples (Table 1). Lateral root growth was severely inhibited to the point that root tips die Reduction in growth may be the consequence of several factors including modification of water status reduction in cell division and cell enlargement, and inactivation of many enzymes by interaction of mercury with -SH groups of the enzymes Data on protein content indicates that at all sampling intervals, the experimentals showed higher levels of protein compared to control. Increased protein content in treated samples points to the inhibitory effect of mercury on the activity of proteolytic enzymes and a reduced degradation of storage protein during germination (Fig. 2). Marked increase in protein content was accompanied by a reduction in growth suggesting that most of the proteins are diverted from a state of growth to a state of osmo-regulation, thus proteins might play a major contribution to osmotic adjustments.

Fig.1 Effect of $HgCl_2$ on seed germination in *V.radiata*

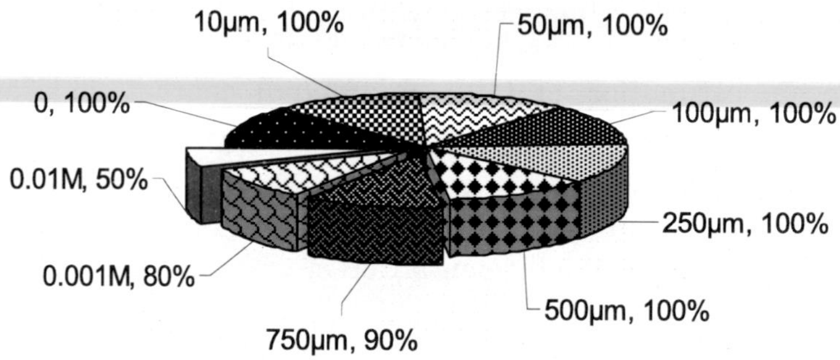

Proline accumulation was shown by all treated samples (Fig. 3). Protection from osmotic injury is accompanied by accumulation of organic osmolytes. Although the precise role of proline accumulation is still debated, proline is considered as a compatible solute involved in osmotic adjustments. Accumulation of proline in mercury stressed seedlings may be due to increased synthesis, concomitant with inhibition in its catabolism Proline accumulation in response to stress has been reported by many workers (Laik *et al.*, 1985 and Machakova *et al.*, 1989).

Table 1 Effect of Mercuric chloride on seedling vigor, root tolerance index and % phytotoxicity in *V.radiata*.

Treatment $HgCl_2$	Seedling vigor	Root tolerance index	% phytotoxicity
Control	1495	100	0
10µm	1090	45.63	54.36
50µm	765	24.75	75.24
100µm	745	24.27	75.73
250µm	480	15.04	84.95
500µm	229	9.12	90.87

Fig.2 Effect of $HgCl_2$ on Protein content in *V.radiata* seeds during germination.

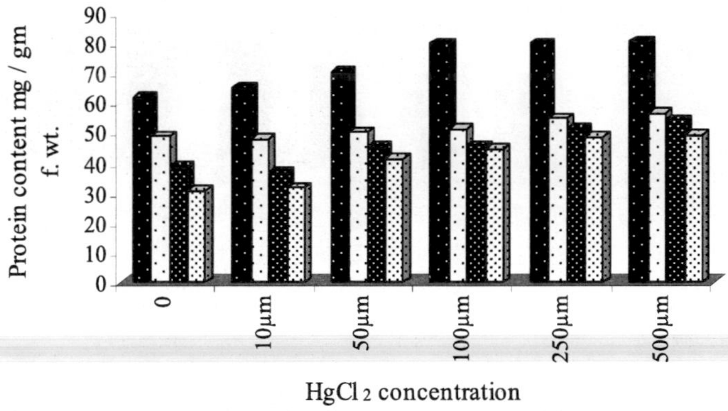

Total Chlorophyll, chlorophyll a & b, leaf area and seedling length decreased with increase in $HgCl_2$ levels on the 7th day (Figs. 4, 5 & 6). Decreased chlorophyll content in $HgCl_2$ treated plants indicates adverse effects of the metal. Photosynthetic efficiency in crop plants is related to the content of chlorophyll in leaves and is regarded as an index of the plants metabolic efficiency to utilize the nutrients absorbed. Zelitch, (1971) reported increase in photosynthetic rate with increase in chlorophyll content of the leaves.

Table 2. Effect of Ascorbic acid in alleviating Mercuric chloride toxicity

Treatment		Germination %
Mercuric chloride	Ascorbic acid	
Control	0	100
750μm	0	90
0.001M	0	80
0.01M	0	50
0	100 μm	100
750μm	100 μm	100
0.001M	100 μm	100
0.01M	100 μm	80

Fig.3 Effect of $HgCl_2$ on Proline content in *V.radiata* (after 96 hrs)

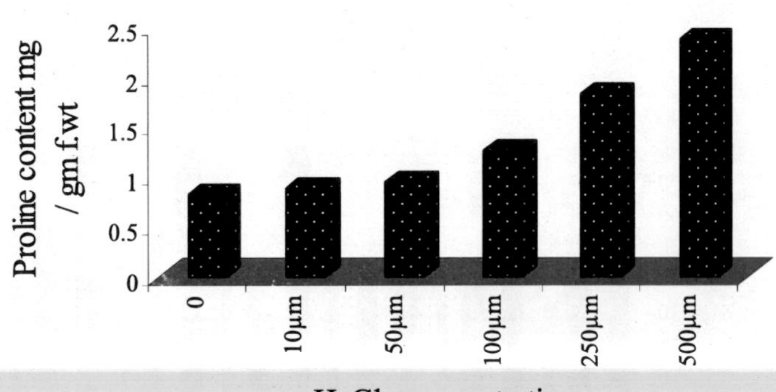

Ascorbic acid in combination with the heavy metal was found to reduce the toxic effect of the metal on germination as recorded by increase in germination % at inhibitory levels of mercuric chloride (Table 2). Antioxidants such as ascorbic acid and œ-tocopherol have been reported to ameliorate the toxic effect of lead and mercury on germination and growth of rice seedlings (Misra and Choudhuri, 1998). Ascorbic acid considered to be a scavenger of free radicals and may counteract membrane lipid peroxidation (Buckland *et al.*, 1991).

Fig.4 Effect of $HgCl_2$ on Chlorophyll content

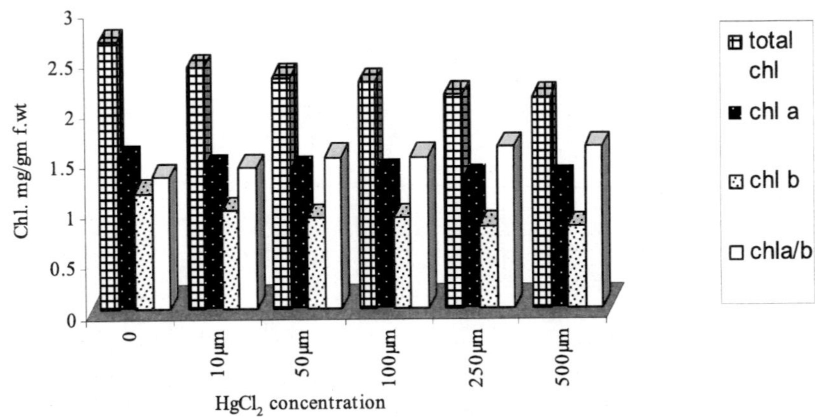

Fig.5 Effect of $HgCl_2$ on Leaf area in *V. radiata*

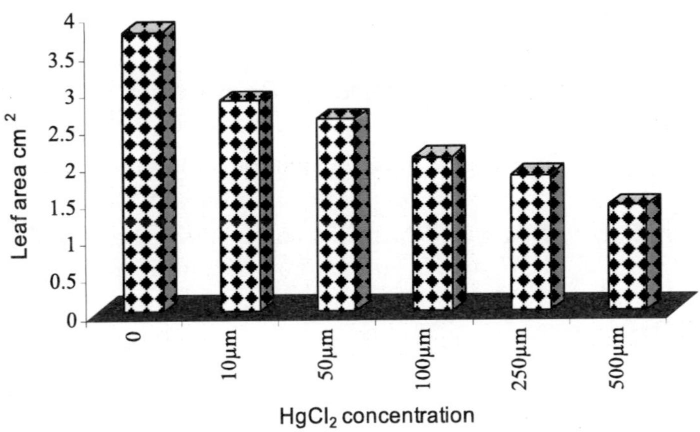

It is concluded that mercuric toxicity in green gram led to decrease in germination %, seedling length, seedling vigor, chlorophyll content and leaf area. Increase in protein content and accumulation of proline occurred in stressed plants. The adverse effect of the metal on germination was reduced by combination with antioxidant ascorbic acid. Increase in global mercury pollution creates the need for increased awareness of scientists and laymen to the consequences leading to disturbances in the ecosystem, and possible harmful effects on plants, animals and man. The need of the hour is immediate identification, monitoring and surveillance of 'hot spots' of metal contamination.

Fig. 6 Effect of mercuric chloride on seedling length (7th)day

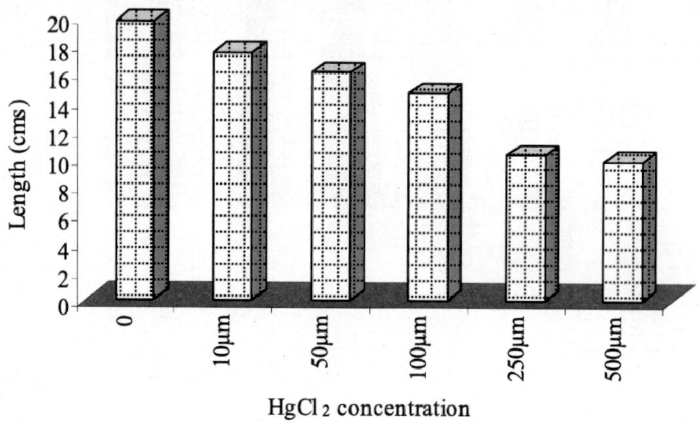

References

Abdul-Baki, A.A and J.D. Anderson 1973. Relationship between decarboxylation of glutamic acid and vigor of soyabean seeds. *Crop Sci.* 13:227-232

Arnon, D.I. 1949. Copper enzymes in isolated chloroplasts. Plant Physiol. 24:1-45.

Buckland, S.M., H.P. Adam and A. F.H. George 1991. The role of ascorbate in drought treated *Cochlearia atlandica* Poped. and *Armeria maritime* (Mill) Willd. New Phytol. 119:155-160.

Bates, L.S., R.P. Waldren and I.D. Teare 1973. Rapid determination of proline for water stress studies. *Plant Soil.* 39:205-207.

Chou, C.H. & C.H. Muller 1972. Allelopatheic mechanism of *Archstaphylous glandulosa var.* Zazaensis. Am. Mid. Nat. 88:324-347.

Laik, I. & K. Dorfling 1985. Hardening, ABA,proline and freezing resistance in two wheat varities. *Physiol. Plant.* 63: 287-292.

Lowry, O.H., N.J. Rosenbrough, A.L. Farr and R.J. Randall 1951. Protein measurements with Folin – Phenol reagent. *J. Biol.Chem.* 193:265-275.

Machakova, I., A. Hansisova and J. Krekule 1989. Levels of ethylene, ACC, MACC, ABA, and proline as an indicator of cold hardening and frost resistance in winter wheat. *Physiol. Plant.* 76: 603-607.

Misra, A. and M.A. Choudhuri 1998. Amelioration of lead and mercury effects on germination and rice seedling growth by antioxidants. *Biol. Plant.* 41:469-473.

Turner, R.C. and C. Marshel. 1972. Accumulation of Zn by sub cellular fraction of root of *Agrostis tenuis* in relation to Zn tolerance. New Phytol. 71: 671-676.

Woodstock, L.W. 1969. Seedling vigor as a measure of seed vigor. Proc. Int. Seed Test. Ass. 34:275-280.

Zelitch, I. 1971. Photosynthesis, Photorespiration and Plant Productivity. Academic Press, New York. p. 347.

ANT-PLANT INTERACTION – IS A FACTOR THAT DETERMINE ANT DIVERSITY?

E. Chellapandi, P. Chandrasekar, S.P. Sevarkodiyone and S. Baskaran

Post Graduate and Research Department of Zoology,
Ayya Nadar Janaki Ammal College, Sivakasi, Tamil nadu - 626 124.

Abstract

Biodiversity entitle the interlinked web of life forms ranging from man to microbes. Equally important, it is a base for continuing evaluation, which is a phenomenon that has enriched our lives. Ants occupy an unique position in insect word in relation to man being beneficial, harmful and competitors of man. Ants are increasingly appreciated as bioindicator for environmental monitoring system. The knowledge of biodiversity of ants is useful for clear understanding of food web, which are important in the functioning of ecosystem. Most of the ants having multi velocity relationship with plants either for getting food from the plant or from the plant - dependant sucking pests or for protecting the plant from natural enemies. In this work, diversity of ants and their relationship with the plants were analysed. Formicinae was a dominant group of ants than other subfamilies. And also, the diversity of ants was positively correlated with rain fall.

Introduction

Biodiversity refers to the variety of life forms, the different plants, animals, insects, and microorganisms, the genes they contain and the ecosystem they form. Biodiversity is defined by the convention of "Biological diversity" (Watson *et al.,* 1995). The insects are the most successful group in the world and they are distributed in all the habitats (Wigglesworth, 1982). Insects are extremely important components of the bioindicators of the world. Insects are interacting with biotic and abiotic factors at different trophic levels in an ecosystem (Chakaravarthy *et al.,* 1997).

The insect diversity contributes to the development of agriculture, medicine and scientific research (Ambrose, 1995). Ants have existed for more than 80 million years and represents one of the most advanced and successful groups of the large order Hymenoptera (Robinson, 1996). The ants are among the most familiar and highly evolved hymenopteran social insects (Krebs, 1985). Ants are increasingly appreciated as bionidicators for environmental monitoring systems (Andersen, 1997, Peck *et al.,* 1998).

Some ants are used as biological control agents, some are in pollination; soil improvement and nutrient cycling (Gotwald, 1986). Some ants get mutualistic partners with other organisms (Cushman and Whitham, 1989). A Knowledge of biodiversity is useful for clear understanding of food webs which are important in the functioning of ecosystems and biodiversity studies in order to conserve organisms and to enable us to adopt to the changing environment (Ananthakrishnan, 1990).

Types of ant-plant interactions

(A) Antagonistic Interactions

Ants have an antagonistic relationship to plants in various ways.

(e.g) Ant-fungus mutualisms

Obligative Associations

Ants and *Acacia* plants for food.

Faculative Associations

Ants and *Macaranga* for food.

Seed Dispersal

Co-evolution is favoured by seed dispersal

(B) Mutualistic Interaction

Association between ant species like *Brachymyrmex* sp., *Camponotus* sp., *Solenopsis* sp., and *Crematogaster* sp. With honey dew producing homopterans on plant foliage.

Materials and methods

Ant species diversity and its association with the selected plants was studied in Neeravipatti village, virudhunagar District. Ants were collected once in a fortnight from the selected plants and analysed.

The main goal of this study was aimed to determine the rate or level of relationship/interaction between the ants and plants on seed predation, mutualism and antagonism in dry and wet seasons.

Methods of collection

Sampling method

All out search method was employed for collecting the ants.

Trapping method

This method was also adopted to collect the flying ants. Four polythene bags (100 ml capacity) were taken. It consists of sugar as attractant. Out of 4 each 2 beakers were tied with the stem or branch portion of the selected plants at one meter height above the ground level using rubber bands and kept it for one day. Then the trapped ants were collected, separated and preserved in 70% alcohol. All ants were identified up to the genus level using the taxonomic keys by B. Holldobler and E.O. Wilson (1990) (Table 1).

Table 1. Diversity of ants in two selected plants of Neeravipatti Village near Sattur during the study period (From August 2003 to February 2004)

S.N	Ants	*Acacia nilotica*							*Azadirachta indica*						
		A	S	O	N	D	J	F	A	S	O	N	D	J	F
	Formicinae														
1	*Camponotus sp.1*	-	-	-	-	-	-	-	-	-	-	+	+	-	-
2	*Camponotus sp.2*	-	-	-	-	-	-	-	+	+	+	+	+	+	+
3	*Camponotus sp.3*	-	-	-	-	-	-	-	-	+	+	-	-	-	-
4	*Crematogaster sp.*	+	+	+	+	+	+	+	+	+	+	-	-	-	-
5	*Anoplolepsis sp.*	+	+	+	-	-	-	-	-	-	-	-	-	-	-
6	*Solenopsis sp.*	-	-	-	-	-	-	-	-	-	-	-	-	-	-
7	*Tetramorium sp.*	-	-	-	-	-	-	-	+	+	+	+	+	+	+
8	*Tapinoma sp.*	-	-	-	-	-	-	-	+	+	+	+	+	+	+

Key : A = August, S = September, O = October, N = November, D = December, J = January, F = February

All-out-search method

In addition to trapping an intensive all-out-search method was also employed to collect the ants.

Preservation of specimen and data recording

The data regarding the total catch in area and also the number of species were recorded in proforma. The data tabulated for measuring various diversity indices such as α-diversity, β-diversity, species richness, evenness indices and relative abundance.

Results

In the present study, the diversity of ants in two different selected plants in Neeravipatti Village near Sattur was analysed. A total number of 8 ant species belonging to3 subfamilies were observed (Table 2, 3, 4, 5 and 6).

Species dominance (Simpson's index)

From the work carried out on the diversity of ants of two different selected plants, the *Camponotus sp.* 1 was evenly distributed in all the plants. The species dominance of different varieties of ants in all two plants was represented as follows:

Crematogaster sp. (ξ = 158.57) was the most abundant species followed by *Tetramorium sp.* (ξ = 42.86), *Tapinoma sp.* (ξ =41.43), *Solanopsis sp.* (ξ =21.43), *Anoplolepis sp* (ξ = 14.29), *Camponotus sp.*1 (ξ = 12.86), *Camponotus sp.*2 (ξ =6.43), *Camponotus sp.*3 (ξ =6.43).

Ant species richness

Out of two different selected plants, ant species richness (R1 and R2) was found to be high in *Azadirachta indica* (ξ =0.6560) and (ξ = 0.3819) than *Acacia nilotica* (ξ =0.2949 and ξ=0.2114).

Diversity indices (Shannon`s index)

The highest mean value of Shannon`s index (H`) of ant diversity was noticed in *Azadirachta indica* (ξ= 1.1874), than *Acacia nilotica* (ξ =0.5844).

Eveness indices/hill`s ratio

In the present investication relative abundance of ant families including Myrmicinae, Formicinae and Dolichoderinae was dominant in *Azadirachta indica* and lower *Acacia nilotica*.

Table 2. Ant diversity in *Acacia nilotica*

S.N	Ant	Sub Family	Aug 2003	Sep 2003	Oct 2003	Nov 2003	Dec 2003	Jan 2004	Feb 2004	ξ
1	*Crematogaster* sp.	Formicinae	100	110	100	500	100	50	50	158.57
2	*Anoplolepis* sp.	Myrmicinae	50	30	20	-	-	-	-	14.29
3	*Solenopsis invicta*	Myrmicinae	50	30	20	10	20	10	10	21.43

Table 3: Ant diversity in *Azadirachta indica*

S.N	Ant	Sub Family	Aug 2003	Sep 2003	Oct 2003	Nov 2003	Dec 2003	Jan 2004	Feb 2004	ξ
1.	*Camponotus sp.1*	Formicinae	10	15	20	10	20	10	5	12.86
2.	*Camponotus sp.2*	Formicinae	-	20	25	-	-	-	-	6.43
3.	*Camponotus sp.3*	Formicinae	10	15	20	-	-	-	-	6.43
4.	*Tetramorium sp*	Myrmicinae	50	35	50	30	50	40	45	42.86
5.	*Tapinoma sp.*	Dolichoderinae	60	40	25	50	50	30	35	41.43

Table 4. Ant richness, diversity and eveness in *Acacia nilotica*

	Aug 2003	Sep 2003	Oct 2003	Nov 2003	Dec 2003	Jan 2004	Feb 2004	ξ
RICHNESS								
NO.	3	3	3	2	2	2	2	2.4286
R1	0.3775	0.3894	0.4047	0.1604	0.2442	0.2442	0.2442	0.2949
R2	0.2121	0.2309	0.2535	0.0088	0.2582	0.2582	0.2582	0.2114
DIVERSITY								
Λ	0.3719	0.4779	0.5478	0.9615	0.7199	0.7175	0.7175	0.6449
H'	1.0397	0.8939	0.7963	0.0096	0.4506	0.4506	0.4506	0.5844
N1	2.8284	2.4446	2.2173	1.1013	1.5692	1.5692	1.5692	1.8827
N2	2.6892	2.0925	1.8255	1.0401	1.3891	1.3891	1.3891	1.6894
EVENNESS								
E1	0.9464	0.8137	0.7248	0.1392	0.6500	0.6500	0.6500	0.6534
E2	0.9428	0.8149	0.7391	0.5507	0.7846	0.7846	0.7846	0.7716
E3	0.9142	0.7223	0.6087	0.1013	0.5692	0.5692	0.5692	0.5792

E4	0.9508	0.8560	0.8233	0.9444	0.8852	0.8852	0.8852	0.8909
E5	0.9238	0.7563	0.6781	0.9554	0.6836	0.6917	0.6917	0.6887

Key:

NO= Number of species

R1= Margalef's index

R2= Menhinick's index

N1= Type-I: Those most sensitive to changes in rare species.

N2= Type-II: Those most sensitive to changes in common species.

Λ = Simpson's index

H' = Shannon's index

E1 to E5 = Modified Hill's ratio

Table 5. Ant richness, diversity and eveness in *Azadirachta indica*

	Aug 2003	Sep 2003	Oct 2003	Nov 2003	Dec 2003	Jan 2004	Feb 2004	ξ
RICHNESS								
NO.	4	5	6	4	4	3	3	4.1429
R1	0.6163	0.8284	0.9529	0.6635	0.6245	0.4564	0.4502	0.6560
R2	0.3508	0.4472	0.4353	0.4170	0.3621	0.3354	0.3254	0.3819
DIVERSITY								
Λ	0.3679	0.2290	0.1910	0.4075	0.3578	0.3987	0.4468	0.3428
H'	1.1190	1.5231	1.7103	1.0213	1.0950	0.9743	0.8687	1.1874
N1	3.0617	4.5865	5.5307	2.7767	2.9891	2.6494	2.3839	3.4254
N2	2.7180	4.3662	5.2347	2.4537	2.7948	2.5079	2.2382	3.1876
EVENNESS								
E1	0.8072	0.9464	0.9545	0.7367	0.7899	0.8869	0.4907	0.8018
E2	0.7654	0.9173	0.9218	0.6943	0.7473	0.8831	0.7946	0.8177
E3	0.6872	0.8966	0.9061	0.5922	0.6630	0.8247	0.6919	0.7571
E4	0.8877	0.9520	0.9465	0.8737	0.9350	0.9464	0.9389	0.9272
E5	0.8333	0.9386	0.9347	0.8182	0.9023	0.9143	0.8948	0.8909

Key

NO. = Number of species

R1 = Margalef's index

R2 = Menhinick's index

N1 = Type-I: Those most sensitive to changes in rare species.

N2 = Type-II: Those most sensitive to changes in common species.

Λ = Simpson's index

H' = Shannon's index

E1 to E5 = Modified Hill's ratio

Table 6. Rainfall during study period from August 2003 to Febraury 2004

Parameters	Aug 2003	Sep 2003	Oct 2003	Nov 2003	Dec 2003	Jan 2004	Feb 2004	ξ
Rainfall	21.2	31.8	119.8	66.4	-	5.2	-	48.88

Discussion

Species diversity indices, such as, species richness and species evenness were studied in order to analyse the association of ants with the two different selected plants and an extensive study was also carried out to know the relationship that exists between the physical parameter and diverty indices.

Species dominance (Simpson's index)

Among the two plants studied, the ant diversity higher in *Azadirachta indica* (7 species), than *Acacia nilotica* (3 species).

Douglas and Sudd (1980) and Robinson (1996) reported that the majority of ant Anoplolepis sp. And Camponotus sp.1 are predaceous or feed on aphid-honey dew. From the observation it was understood that the numerical superiority of *Crematogaster sp., Anoplolepis sp., Camponotus sp.1* was more in both plants. This is because of their close association with hemipteran pest, a well known vectors of plant diseases. (Kenne, *et al.*, 2003).

In the present investigation, it was found that the number of *Crematogaster* sp. was rich in *Acacia nilotica.* This may be due to mutualistic ant – plant association, a common phenomenon in many ecosystem, Here *Acacia nilotica* (Myrmecophytes) provides nesting space (domatia) and often also food for ants. In exchange, the ant inhabitans affectively protect their host plant against herbivory.

Janzen (1973) Feldhaar *et al.* (2003) reported that the colonization of *Crematogaster* sp. was positively increased with the growth of host *Macaranga (*Euphorbiaceae).

Ant species richess

The ant species richness (R1 and R2) was found to be high in *Azadirachta indica* than other one. The main reason was the occurrence of facultative association of *Camponotus* sp., *Anoplolepis* sp., *Tapinoma* sp., with *Azadirachta indica* since these ants are considered as seed dispersers during fruiting season (Gomez *et al.*, 2003) and Fleming and Nicolson, 2003.)

Diversity indices (Shannon's index)

The highest Shannon's index (H') in the ant diversity was noticed in *Azadirachta indica* than *Acacia nilotica.* The value of Shannon's index (H'), the greater the species diversity of ant community (Dash, 1996).

Evenness indices/hill's ratio

In this study all the species of different ant sub families were evenly distributed in both the selected plants. The reason for the even distribution of ant species is due to the influence of inter and intraspecific competition (Yu and Davidson, 1997); Chemicalcues, or habitat preferences of both partners (Inui *et al.*, 2001).

Relative abundance

The relative abundance of individuals of family Formicinae was more in *Azadirachta indica.* The reason for the over population of subfamily was mainly due to the feeding habit on honey dews of homopterans and domatia, the microhabitat for ants (Stuntz *et al* 2002).

Palanichamy *et al.* (1995), also reported that black ant *Camponous compressus* was the major pollinators of flowering plants.

Physical parameter

Rainfall is one of the most important factors in an environment since they interfere with the growth and distribution of flora and fauna (Hutchinson, 1967). In the present investigation the ant diversity was positively correlated with rainfall.

Acknowledgement

The authors are thankful to the management and principal for providing facilities

References

Ambrose, D.P. 1995. Insect diversity and conservation, *Zoo's, print,* X (10): 9-11

Andersen, A. 1997. Using ants as a bioindicators,*conservation ecology,* 1:8.

Chakaravarthy, A.K., Rajagopal, D. and R. Jeganatha 1997. Insects as bioindicators of conservation in the tropics, *Zoo's print,* **XII** (1): 21-25.

Cushman, J.H and T.G. Whitham 1989. Conditional interactions in ant-plant-herbivore mutualisms. In: *Ant-Plant Interaction,* Huxley,C.R and D.F. Cutler [eds.] Oxford University, Oxford, 92-103.

Douglas, J.M. and J.H. Sudd 1980. Behavioural coordination between an aphid (*Symydobious oblongus*) and the ant that attends it (*Formica Lugubris*) : An ethological analysis, *Anim. Behavi.,* **28** (4) : 1127-1139.

Fedlhaar, H., Fiala, B., Hashim, R.B. and U. Maschwitz 2003. Patterns of the *Crematogaster – Macaranga* association : The ant partner makes the difference, *Insectes Soc.,* **50** : 9-19.

Fleming, P.A and S.W. Nicolson 2003. Arthropod fauna of mammal-pollinated *Protea humiflora:* Ants as attractant for insectivore pollinators? *African. Entomol.,* **11** (1) : 9-14.

Gomez, C., Pons, P and M.B. Joseph 2003. Effects of the Argentine and *Line pithema humile* on seed dispersal and seedling emergenece of *Rhamnus alaternus, Ecography,* **26** (4) : 532-538.

Gotwald, W.H. 1986. The beneficial and economic role of ants In : B (Ed. S.B. Vinson), Praeger, New York, 290-313.

Holldobler, B. and E.O., Wilson 1990. *The Ants,* Harvard University Press, Cambridge, Mass : 732.

Hutchinson, G.E. 1967. Introduction to lake biology and the Limnoplankton. John Wiley and Sons Inc., New York, *Areattise on Limnology,* **2.**

Inui; Y., Itioka, T., Murase, K., Yamaoka, P. and T. Itino 2001. Chemical recognition of partner plant species by founders ant queens in Mecavanga. (Vematogaotes myrmecophytism) *J. Chem. Ecol.,* 27: 2029-2040.

Janzen, H.D. 1973. Evolution of polygyrous obligate *Acacia*-ants in Western Mexico. J. *Anim. Ecol.,* **42 :** 727-780.

Kenne, M., Djie. T., Lordon. C., Orivel, J., Mony, R., Fabre, A. and A. Dejean 2003. Influence of insecticide treatment on Ant Hemiptera associations in tropical plantations, *J. Econ. Entomol.,* **96** (2) : 251-258.

Krebs, C.J. 1985. *Ecology – The Experimental Analysis of Distribution and Abudance.* Horper and Row, New York, 37-39.

Nair. M.R.G.K. 1986. *Insects and Mites of Crops in India ICAR,* New Delhi; 408.

Palanichamy, P., Baskaran, S and A. Mohandoss 1995. Insects pollination of the moringa plant, *Moringa concanensis inimmo Linn, Environ, Ecol.,* **13** (1) : 45-51.

Robinson, W. 1996. *Urban Entomology,* Chapman and Hall, London, 262-284.

Stuntz, S., Ziegler, C., Simon, U. and G. Gerhard 2002. Diversity and structure of the arthropod fauna within three canopy epiphyte species in Central Panama. Cambridge University Press. *J. Tropical Ecology,* **18** 161-176

Watson, R.T., Heywood, V.H., Baste., I., Dias, B. and G. Ruarks (Eds.) 1995. Global biodiversity assessment for policy makers, Publishers for the United Nations Environment Programme, Cambridge University Press, Cambridge, **46.**

Wigglesworth, V.B. 1982. *The Principles of Insect Physiology.* Seventh Edition, Chapman and Hall, London.

Yu, D.W and D.W. Daridson 1997. Experimental studies of species specificity in Cecropia and relationship. *Ecol. Mongor.,* 67: 273-294.